Synthesis Lectures on Chemical Engineering and Biochemical Engineering

This series publishes short books on all aspects of chemical engineering, covering the analysis or design of chemical processes to effectively convert materials into more useful materials or energy. The books will focus on fundamental aspects necessary for chemical engineering design including chemistry, math, physics, and sometimes biology to improve the quality of life by inventing, optimizing, and economizing new technologies and products.

Subir Bhattacharjee · Jimmy Yu

Digital Transformation of the Chemical Process Industry

A Practical Guide for Professionals

Subir Bhattacharjee
IntelliFlux Controls, Inc.
Irvine, CA, USA

Jimmy Yu
Water Sustainability and Processing
Technology
Global R&D, Pepsico
Valhalla, NY, USA

ISSN 2327-6738   ISSN 2327-6746 (electronic)
Synthesis Lectures on Chemical Engineering and Biochemical Engineering
ISBN 978-3-031-94053-8   ISBN 978-3-031-94054-5 (eBook)
https://doi.org/10.1007/978-3-031-94054-5

© The Editor(s) (if applicable) and The Author(s), under exclusive license to Springer
Nature Switzerland AG 2026

This work is subject to copyright. All rights are solely and exclusively licensed by the Publisher, whether the whole or part of the material is concerned, specifically the rights of translation, reprinting, reuse of illustrations, recitation, broadcasting, reproduction on microfilms or in any other physical way, and transmission or information storage and retrieval, electronic adaptation, computer software, or by similar or dissimilar methodology now known or hereafter developed.
The use of general descriptive names, registered names, trademarks, service marks, etc. in this publication does not imply, even in the absence of a specific statement, that such names are exempt from the relevant protective laws and regulations and therefore free for general use.
The publisher, the authors and the editors are safe to assume that the advice and information in this book are believed to be true and accurate at the date of publication. Neither the publisher nor the authors or the editors give a warranty, expressed or implied, with respect to the material contained herein or for any errors or omissions that may have been made. The publisher remains neutral with regard to jurisdictional claims in published maps and institutional affiliations.

This Springer imprint is published by the registered company Springer Nature Switzerland AG
The registered company address is: Gewerbestrasse 11, 6330 Cham, Switzerland

If disposing of this product, please recycle the paper.

Preface

Welcome to the realm of digital transformation in the chemical process industry—a journey that promises to reshape the very foundations of how we conceive, design, and operate industrial chemical processes. In the dynamic landscape of today's industrial sector, where innovation, agility, and adaptability are of paramount importance, embracing the possibilities offered by digital technologies is not just a choice; it is a strategic imperative. This book serves as a practical instruction manual to guide the reader through the intricate maze of digital transformation, specifically tailored for the nuances of the chemical process industry. As we stand at the intersection of traditional methodologies and cutting-edge technologies, the need for a comprehensive understanding of the digital ecosystem has never been more pressing for the process engineer and practitioner. The chemical sector, with its complex processes and supply chains, is uniquely positioned to unlock unprecedented value through the intelligent use of digital tools.

This book targets technical professionals and business leadership regarding implementation of digital transformation in their organizations with a particular focus on implementation of digitalization in the chemical processing industry (CPI). The book focuses on the preliminary steps in planning the critical organizational, behavioral, cultural, and technical nuances of digital transformation in an enterprise, and provides a roadmap of how to implement these steps. This evolutionary process is mapped out and navigated with experiential learnings from multiple segments of the process industry. The main thesis of the book is that digital transformation is equally and perhaps more importantly dominated by cultural and human aspects as today's process environments represent "human in the loop" systems. Implementation of digital solutions will need an amalgamation of information technology and data science with human collaboration, workflows, and organizational cultures. The book will provide practical guidelines of implementing these human-machine interfaces across critical operational and management facets of businesses.

Should a comprehensive book on the digital transformation of the chemical processing industry (CPI) be written? And if so, what should its content encompass? These questions have long been central to discussions within the industry, particularly as companies strive to modernize their operations in an era of rapid technological advancement. While extensive literature exists on digital transformation in manufacturing—often focusing on discrete production and robotic automation—guidance specific to the process industries remains fragmented. This book seeks to address that gap, providing a structured roadmap for digitalization in CPI, with practical insights drawn from real-world implementation.

Over the past decade, significant efforts have been dedicated to developing and implementing digital systems for resource management, particularly energy and raw material minimization, water treatment for reuse or recycle, and waste stream management. Asset management, process automation, remote monitoring, and autonomous control with a focus on maximizing capacity utilization have also been major initiatives within the CPI. These efforts have generated vast amounts of data, requiring robust analytics to extract meaningful insights for optimizing operations. Many companies within the CPI are eager to embrace digital transformation, recognizing the potential of Industrie 4.0 (I4.0), the Internet of Things (IoT), smart factories, and cyber-physical systems (CPS). However, despite this enthusiasm, challenges persist in translating these broad concepts into practical, plant-level strategies. Questions regarding prioritization, planning, implementation, and investment remain largely unanswered, and this book aims to uncover some of these answers.

This book is the outcome of extensive fieldwork and research in digital transformation, with a particular focus on the process industry. Drawing from a structured, multi-year collaboration that explored digitalization in process-intensive sectors such as food and beverage production, oil and gas, and water purification, the insights presented here offer an objective and experience-based perspective on the topic. Unlike many existing resources that primarily provide theoretical overviews drawing broadly from generalizations and abstractions originating from the information management background, this work emphasizes real-world applications specific to the chemical processing industry, documenting implementation strategies, tracking progress, and defining success metrics for digital adoption in the CPI.

Digital transformation is not just about technology—it can fundamentally shift how industries operate, requiring a seamless integration of automation, augmentation, and human insight. Too often, discussions focus on technical implementation while overlooking the human behaviors and cultural shifts necessary for success. A human-centric approach ensures that digitalization enhances decision-making, improves efficiency, and drives meaningful change rather than simply replacing old processes with new technology. This is what we hope to explore in this book in addition to the technical discussions, shedding light on how to achieve a balanced, effective transformation.

The rapid pace of technological advancement has outstripped traditional models of adaptation, creating a gap that must be addressed through strategic integration rather than just upskilling. Automation is most powerful when it augments human expertise, enabling organizations to harness data, optimize operations, and drive innovation. True digital transformation happens when technology and people work in harmony, creating a resilient, adaptive, and forward-thinking organization ready to shape the future.

The book is structured into three parts. Part I outlines the rationale for pursuing digital transformation in the chemical process industry (CPI). It begins with an overarching view of industry ambitions and goals, followed by an exploration of how digital technologies are reshaping business and technical perspectives to achieve these objectives. This part concludes with a discussion of Industrie 4.0 design principles in manufacturing. Part II focuses on the technical implementation of digital transformation in the CPI, with an emphasis on retrofitting existing chemical plants. Chapters 5–7 cover the planning phase, baseline implementation, and advanced strategies for digital transformation. Finally, Part III (Chaps. 8–10) examines the organizational, human, and cultural dimensions of digital transformation, considering their impact in the evolving Industrie 5.0 landscape.

This book serves as playbook and a guiding principle for CPI professionals navigating digital transformation. It is designed to provide actionable insights, real-world examples, and a forward-looking perspective on the future of digitalization in process industries. The transformation of CPI is not a distant vision; it is already underway. Those who embrace this shift will not only adapt but will also lead the next industrial revolution.

Let the journey begin.

Irvine, USA Subir Bhattacharjee
Valhalla, USA Jimmy Yu

Acknowledgments Writing this book has been an extraordinary journey, made possible by the support, guidance, and contributions of many individuals and organizations. We would like to express our sincere appreciation to our colleagues, whose insights, feedback, and expertise have greatly influenced our understanding of digital transformation.

Much of my (SB) understanding of digitalization in the process industry developed during my tenure at IntelliFlux Controls. I am especially grateful to my team members—Matt McCallum, Wenxin Hu, Xueying Dai, Anthony Valverde, and Gil Hurwitz—who brought my abstract ideas on digital transformation to life across multiple plants worldwide. I would also like to take this opportunity to thank Eric Hoek, Arian Edalat, Richard Wolfen, Anthony Wachinski, Gurudev Singh, Emrah Ercan, Shawn M. Steele, Peter Fiske, Bryan Terrell, and Hasit Joshipura for their invaluable support, guidance, and collaboration. Additionally, I acknowledge the support of organizations such as ImagineH2O and PUB, Singapore, whose contributions during IntelliFlux's formative years were instrumental in shaping my journey.

I (JY) am incredibly grateful to Wenny Noha as a thought partner who constantly challenges me and reshapes my thinking on this topic. I also extend my deep appreciation to Anthony Weishampel and Jason Parcon for teaching me the importance of statistics and the fundamentals of machine learning, and to Mina Sfondilis, whose insights as a behavioral scientist have opened my eyes to the critical role of human-centricity in this transformation. Our journey (SB and JY) in digital transformation has been further enriched by Andry Agustiady and Brian Boothe, who continue to expand our perspective, especially on the operation side.

We are grateful to our publisher, particularly Arza Seidel, for dealing with our numerous requests, delays, and interminable revisions and edits of the book with grace and patience.

We thank our families for their unwavering encouragement and patience.

Competing Interests The authors have no competing interests to declare that are relevant to the content of this manuscript.

Authors' Note

The views, opinions, and ideas expressed in this book are solely those of the authors and do not represent, reflect, or imply the positions, policies, endorsements, or official perspectives of any organization with which the authors are affiliated, including but not limited to IntelliFlux Controls, Inc. (for SB) and PepsiCo, Inc. (for JY). Furthermore, no part of this book should be construed as being endorsed by, attributed to, or influenced by any of the authors' funding sources, supporting organizations, or employers. The authors assume full responsibility for the content, analysis, and interpretations presented herein. Neither the authors nor their affiliated organizations shall be held liable for any consequences arising from the use, application, or reliance upon the material contained in this book.

The mention of any commercial product, service, or company in this book is for informational purposes only and does not constitute an endorsement, sponsorship, or recommendation by the authors. The authors have no financial interest in, nor do they receive any compensation or other benefits from, the inclusion of such references. Furthermore, the authors make no representations or warranties regarding the performance, suitability, or reliability of any mentioned products, services, or companies. Readers are solely responsible for conducting their own due diligence and exercising independent judgment before making any decisions based on the information presented in this book. The authors assume no liability for any consequences resulting from the use of or reliance on such information.

The authors acknowledge the use of ChatGPT as an editorial aid in the writing and refinement of this book. ChatGPT was utilized for tasks such as drafting, revising, and enhancing the clarity and coherence of certain sections. While the content, ideas, and conclusions presented in this work are solely those of the authors, ChatGPT provided linguistic and stylistic assistance in shaping the final manuscript. Any errors or omissions remain the responsibility of the authors.

Contents

Part I The Case for Digital Transformation of Chemical Process Industry

1 Introduction 3
   1.1 The Reader of this Book 3
   1.2 The Chemical Processing Industry 5
   1.3 Resource Intensity of Process Industries 6
   1.4 Circular Economy Practices 8
   1.5 Digital Transformation in the Chemical Industry 9
   References 12

2 The Business Case for Digital Transformation 15
   2.1 Introduction 15
   2.2 Chemical Process Industry Paradigms 15
      2.2.1 Sustainability 15
      2.2.2 Reducing Carbon Footprints 16
      2.2.3 Circular Economy and Resource Efficiency 16
      2.2.4 Green Chemistry and Safer Alternatives 16
      2.2.5 Water and Energy Conservation 16
      2.2.6 Transparency and Responsible Sourcing 17
      2.2.7 Regulatory Compliance and Industry Collaboration 17
      2.2.8 Challenges and Future Outlook 17
   2.3 Human Aspects 18
      2.3.1 Human-in-the-Loop Systems 18
      2.3.2 Adaptive Learning and Continuous Improvement 19
      2.3.3 Enhancing User Experience and Trust 19
      2.3.4 Domain-Specific Considerations 19
      2.3.5 Ethical and Social Implications 19
      2.3.6 Role of Digital Transformation in HITL Systems 20
      2.3.7 Challenges and Opportunities 20

	2.4	Process Industry and Information Technology	20
		2.4.1 Enhancing Efficiency Through Automation	21
		2.4.2 Data Analytics and Predictive Maintenance	22
		2.4.3 Integration of Industrial Internet of Things (IIoT)	22
		2.4.4 Cybersecurity Challenges and Solutions	22
	2.5	Digital Transformation Scenarios in Selected Sectors	23
		2.5.1 Petroleum Refining and Petrochemicals Processing	23
		2.5.2 Food and Beverage Industry	24
		2.5.3 Speciality Chemical and Biochemicals Processing	25
		2.5.4 Mining and Mineral Processing	26
		2.5.5 Pharmaceutical Industry	26
	2.6	Agile Business Powered by Digital Transformation	27
		2.6.1 Digital Transformation: Unleashing the Power of Data	27
		2.6.2 Automation: Bridging the Gap Between IT and OT	27
		2.6.3 Sustainability: Balancing Environmental and Business Needs	28
	2.7	Building a Strong Business Case	28
		2.7.1 Navigating Synergy	28
		2.7.2 Catalyzing Operational Excellence	29
		2.7.3 Quality and Innovation Leadership	29
		2.7.4 Supply Chain Resilience and Responsiveness	29
		2.7.5 Safety, Compliance, and Reputation Enhancement	30
		2.7.6 Holistic Sustainability	30
		2.7.7 Customer-Centric Innovation	30
	2.8	Closing Thoughts	31
		Reference	31
3	**The Path to Digital Transformation**		**33**
	3.1	Evolution of Manufacturing	33
	3.2	Progress of Automation in Manufacturing	34
	3.3	Challenges and Opportunities	35
	3.4	The Case for Digital Transformation	36
		3.4.1 Operational Efficiency and Optimization	37
		3.4.2 Predictive Maintenance and Asset Management	37
		3.4.3 Supply Chain Visibility and Collaboration	37
		3.4.4 Quality Control and Compliance	37
		3.4.5 Innovation and Product Development	38
		3.4.6 Energy Efficiency and Sustainability	38
		3.4.7 Risk Management	38
	3.5	Is Digital Transformation a Revolution in the CPI?	39

	3.6	Rationale and Timeline of Implementing Digitalization	40
		3.6.1 Process Plant Life Cycle and Digitalization	40
		3.6.2 Pre-commissioning Digitalization	40
		3.6.3 Retrofit or Post-commissioning Digitalization	43
	3.7	Outlook	44
		Reference	45
4	**Digital Transformation–Technical Foundations**		**47**
	4.1	Introduction	47
	4.2	Technical Premise of Digital Transformation	48
	4.3	The Industrie 4.0 Framework	49
		4.3.1 The PDP Loop	51
		4.3.2 The DIKW Loop	52
	4.4	Components of Digital Transformation	55
		4.4.1 Industrial Control System (ICS)	55
		4.4.2 Information Technology (IT)	58
		4.4.3 The Internet of Things (IoT)	59
		4.4.4 Database as a Single Source of Truth	60
		4.4.5 Computation for Digital Transformation	61
	4.5	Closing Remarks	63
		References	64

Part II The Journey Toward Digital Transformation

5	**Planning Digital Transformation**		**67**
	5.1	Implementing Digitalization—The Roadmap	67
	5.2	Concept Development and Planning	69
		5.2.1 Planning Digital Transformation for an Enterprise	70
		5.2.2 Key Considerations for Digital Transformation Success	70
		5.2.3 Tools for Planning and Execution	71
		5.2.4 Workforce Training and Change Management	73
		5.2.5 Timeline and Continuous Improvement Strategy	73
	5.3	Digitization: Foundation of Digitalization	74
		5.3.1 Key Information for Digitization	76
		5.3.2 Data Structure Coordination Across Digital Platforms	79
		5.3.3 Digitized Data Quality in Process Automation	80
	5.4	The Digitalization Implementation Team	82
	5.5	OT/IT Integration Assessment	83
		5.5.1 Integration Considerations	85
		5.5.2 Hardware and Software Considerations	85
		5.5.3 Integration of IoT Devices in the OT Layer	87

	5.6	Internet of Things	87
		5.6.1 Communication Protocols	90
		5.6.2 Cybersecurity and Authentication Protocols	92
	5.7	Data Structures, Algorithms, and Automation	94
		5.7.1 Computational Philosophy and Infrastructure for Digital Transformation	95
	5.8	Human-Machine Interfaces in a Digitally Transformed Enterprise	98
		5.8.1 The Cyber-Physical System Paradigm	99
		5.8.2 Shifts in Human-Machine Interaction	99
		5.8.3 Decision Support and Operational Setpoints	100
		5.8.4 Business Stakeholder Interactions with Digital Tools	100
		5.8.5 Eliminating Manual Historical Analysis and Enabling Continuous Computing	100
	5.9	Assessment Questions	101
	References		103
6	**Implementing Digital Transformation**		**105**
	6.1	Digital Transformation: From Strategy to Execution	105
	6.2	Pilot Implementation in a Mid-Life Manufacturing Unit	106
		6.2.1 Business Case for Retrofit Digital Transformation in Brownfield Plants	106
		6.2.2 Digital Transformation Readiness Assessment	109
		6.2.3 Strategic Roadmap for Implementation	113
	6.3	Workflow Automation: The Lowest Hanging Fruit in Digital Transformation	114
		6.3.1 Workflow Automation and the DIKW Framework	114
		6.3.2 Examples of Workflow Automation in Process Plants	115
		6.3.3 Key Benefits of Workflow Automation	115
	6.4	Enhancing Operations Through Transparency and Interoperability	116
		6.4.1 Database Development and Analytics Integration	118
		6.4.2 Enhancing Plant Operations with Dynamic Recovery Optimization	120
		6.4.3 Leveraging Transparency and Interoperability for Improved Performance	120
	6.5	Evolution and Maintenance of a Smart Infrastructure	121
		6.5.1 Connectivity and Integration: Building the Digital Thread	121
		6.5.2 Data Analytics: Extracting Actionable Insights	122
		6.5.3 Automation and Robotics: Enhancing Operational Efficiency	122
		6.5.4 Human-in-the-Loop Automation: Augmenting Human Capabilities	122

	6.6	The Challenges of Attaining a Fully Implemented Digitalization Vision	123
		6.6.1 Overcoming Complexity and Integration Challenges	123
		6.6.2 Data Quality and Reliability: A Cornerstone for Decision Automation	123
		6.6.3 Cybersecurity: Safeguarding Against Threats	124
		6.6.4 Job Displacement: Addressing Workforce Concerns	124
		6.6.5 Human Consequences: Balancing Efficiency with Ethical Considerations	124
		6.6.6 Reskilling and Retraining: Investing in Human Capital	124
		6.6.7 Change Management: Cultivating a Culture of Adaptability	125
7	**Integrating Digitalization into the Process Industry**		**127**
	7.1	Introduction	127
	7.2	From Stimulus to Data: Principles of Transduction	128
		7.2.1 Field Instruments	129
		7.2.2 Sensors for Process Monitoring	130
		7.2.3 Workflow and Business Considerations for Sensor Integration	137
	7.3	From Data to Process Information	138
		7.3.1 Thermodynamic Information	138
		7.3.2 Transport Phenomena Modeling	140
		7.3.3 Chemical Reaction Kinetics and Thermochemistry	140
		7.3.4 Physics-Based Digital Twin Modeling of Process Components	141
		7.3.5 Network Modeling of Process Components	143
	7.4	Dimensionality Reduction Using Dimensionless Numbers	146
	7.5	Hybrid Data- and Physics-Driven Approaches	147
	References		149

Part III Corporate Vision and Human Aspects of Digital Transformation

8	**Digital Transformation–Corporate Vision**		**153**
	8.1	Toward Industrie 5.0	153
		8.1.1 Human-Centricity: Redefining the Role of Technology for the Workforce	153
		8.1.2 Moving from Task Automation to Decision Support	154
		8.1.3 Redefining Job Roles and Skills	154
		8.1.4 Enhancing Worker Safety and Well-Being	155

8.2	Sustainability: Expanding the Scope of Industrial Responsibility		155
	8.2.1	Adopting a Holistic Approach to Resource Management	155
	8.2.2	Embracing Circularity Beyond the Plant Boundary	156
	8.2.3	Leveraging Data for Predictive Sustainability	156
8.3	Resilience: Building Robust Systems for an Uncertain Future		156
	8.3.1	Redesigning Systems with Flexibility and Adaptability	157
	8.3.2	Enhancing Supply Chain Resilience Through Digital Connectivity	157
	8.3.3	Fostering a Resilient Workforce Through Cross-Training and Knowledge Sharing	157
8.4	Digitalization: More than CapEX Approach		157
8.5	Digital Transformation: Implementing the Human Element		160
	8.5.1	The Corporate Goal and the Importance of Communication	160
	8.5.2	Three Key Personas in Digital Transformation	161
8.6	Strategy Setters: Crafting an Actionable Plan for Digital Transformation		162
	8.6.1	Aligning Transformation with Corporate Strategy	162
	8.6.2	Crafting a Comprehensive and Scalable Strategy for Implementation	162
	8.6.3	Balancing Business and Technical Considerations	163
	8.6.4	Vendor Selection and Technological Advancements	163
	8.6.5	The Role of Sponsors and Senior Decision-Makers	163
	8.6.6	The Role of Communication in Sustaining Momentum	164
	8.6.7	Ensuring Continuity in Strategy Teams	164
8.7	Project Implementers: Building Cross-Functional Collaboration		164
	8.7.1	Clear Corporate Goals to Successful Implementation	164
	8.7.2	IT-OT Integration: A Cornerstone of Digital Transformation	165
	8.7.3	Cross-Functional Collaboration: Beyond IT and OT	165
	8.7.4	Agile Project Management	165
	8.7.5	Managing Legacy Systems and Technical Debt	166
8.8	Frontline Employee Involvement: A Key to Success		166
	8.8.1	The Importance of Frontline Feedback	166
	8.8.2	Training for Successful Adoption	167
	8.8.3	Empowering Frontline Champions	167
	8.8.4	Providing Ongoing Support	167
	8.8.5	Ensuring Security and Access Control	168
References			168

9	**Human Involvement and Roles**	169
	9.1 Automation and Augmentation for Frontline Employee	169
	9.1.1 Automation: Improving Workflow Efficiency	169
	9.1.2 Augmentation: Empowering the Human Element in Manufacturing	170
	9.1.3 Designing Systems that Balance Automation and Augmentation	171
	9.1.4 The Human-Centric Future of Manufacturing	172
	9.2 Designing a System for Automation and Augmentation: A High-Level Guide	172
	9.2.1 Define the Goals and Identify Use Cases	172
	9.2.2 Map Current Workflows and Processes	173
	9.2.3 Prioritize Data Centralization and Integration	173
	9.2.4 Balance Automation and Human Judgment	174
	9.2.5 Design User-Centered Interfaces and Tools	174
	9.3 Designing Systems to Capture Continuous Feedback	175
	9.3.1 Embedding Feedback in Digital Systems	175
	9.3.2 Evolving Systems Based on User Feedback	176
	9.3.3 Encouraging Continuous Engagement	176
	9.4 Different Persona for Implementation	177
	9.4.1 Design Phase: Subject Matter Experts (SMEs)	177
	9.4.2 Testing Phase: Detail-Oriented Evaluators	178
	9.4.3 Soft Launch: Early Adopters and Technology Enthusiasts	179
	9.4.4 Leadership Champion: Advocates for Transformation	179
	9.5 Conclusion	180
	References	180
10	**Digital Transformation and Cultural Transformation**	181
	10.1 The Human Factor in Digital Transformation–Thoughts	181
	10.2 Defining Culture in a Manufacturing Context	181
	10.2.1 Divergence Between Corporate and Plant Levels	181
	10.2.2 Corporate and Plant Artifacts	182
	10.3 Digital Transformation as a Culture Shaper	184
	10.4 Building Norms Rather than Isolated Cultural Elements	185
	10.4.1 The Peril of Isolated Cultural Events	185
	10.4.2 Building New Norms Through Digital Transformation	186
	10.5 Recognizing and Addressing Cultural Resistance	187
	10.6 Fostering Cultural Shifts Through Digitalization	189
	10.6.1 Increased Observability and Decreased Plausible Deniability	189
	10.6.2 Using Data to Challenge Bias and Drive Decision-Making	190

		10.6.3	Encouraging a Culture of Scientific Debate and Psychological Safety	191
		10.6.4	Shifting Norms Through Data-Driven Discussions	192
		10.6.5	Promoting Collaboration Through Data Sharing	192
		10.6.6	Using Digital Tools to Train and Build Capabilities	193
		10.6.7	Enhancing Corporate Visibility with Digital Data	194
	10.7	Conclusion		195
	References			195
11	**Epilogue–Outlook**			197
	11.1	Fundamental Strategic Goals		197
		11.1.1	Experimentation and Evaluation	197
		11.1.2	Building an Ecosystem	197
		11.1.3	Scaling at the Edges	198
		11.1.4	Proving with Strategic Implementations	198
		11.1.5	Iterative Improvement	198
	11.2	Common Misconceptions and Hyper-Expectations		198
	11.3	Shaping the Future		199

Abbreviations

AI	Artificial Intelligence
API	Application Programming Interface
AR	Augmented Reality
CAN	Controller Area Network
CMMS	Computerized Maintenance Management System
CPI	Chemical Process Industry
CPS	Cyber-Physical System
CRM	Customer Relationship Management
DA	Decision Automation
DAS	Decision Automation System
DCS	Distributed Control System
DIKW	Data, Information, Knowledge, Wisdom
DSS	Decision Support System
DT	Digital Twin
ERP	Enterprise Resource Planning
HART	Highway Addressable Remote Transducer Protocol
HITL	Human-in-the-Loop
HTTP	Hyper-Text Transfer Protocol
ICS	Industrial Control System
IIoT	Industrial Internet of Things
IoT	Internet of Things
IP	Internet Protocol
IT	Information Technology
JSON	Java Script Object Notation
KMS	Knowledge Management System
LED	Light Emitting Diode
LIMS	Laboratory Information Management System
MES	Manufacturing Execution System

ML	Machine Learning
MQTT	Message Queuing Telemetry Transport
OPC	Open Platform Communications
OPC/UA	Open Platform Communications/Unified Architecture
OT	Operational Technology
PLC	Programmable Logic Controller
REST	Representational State Transfer
RPA	Robotic Process Automation
RTOS	Real Time Operating Systems
SCADA	Supervisory Control and Data Acquisition
SCM	Supply Chain Management
SQL	Sequential Query Language
TCP	Transmission Control Protocol
UI	User Interface
UX	User Experience
VFD	Variable Frequency Drive
VR	Virtual Reality

Symbols

α	Thermal diffusivity, or heat transfer coefficient in Fourier's law
ΔG	Gibbs free energy change (J/mol)
ΔH	Enthalpy change (J/mol)
λ	Thermal conductivity (W/m·K)
μ	Dynamic viscosity (Pa·s)
ν	Kinematic viscosity ($\nu = \frac{\mu}{\rho}$)
ρ	Fluid density (kg/m^3)
τ	Time constant of a sensor
Bi	Biot number ($\frac{hL}{k}$), ratio of internal to surface thermal resistance
c_p	Specific heat capacity at constant pressure (J/kg·K)
c_v	Specific heat capacity at constant volume (J/kg·K)
D	Diffusivity or mass diffusion coefficient (m^2/s)
Da	Damköhler number ($\frac{kL}{u}$), compares reaction rate to transport rate
Fo	Fourier number ($\frac{\alpha t}{L^2}$), governing transient heat conduction
K	Proportional gain of a sensor
k	Thermal conductivity or reaction rate constant (depending on context)
m_i	Mass fraction of component i in a mixture
Nu	Nusselt number ($\frac{hL}{k}$), ratio of convective to conductive heat transfer
p	Pressure (Pa)
Pe	Péclet number (Re · Pr), relative importance of advection to diffusion
Pr	Prandtl number $\frac{c_p \mu}{k}$, ratio of momentum to thermal diffusivity
Q	Flow rate, typically Mass (kg/s) but can also be volume (m^3/s)
Q_{in}	Input quantity measured as input signal by a sensor
Re	Reynolds number $\frac{\rho u L}{\mu}$, ratio of inertial to viscous forces
S	Sensitivity of a sensor
Sc	Schmidt number $\frac{\mu}{\rho D}$, ratio of momentum to mass diffusivity

Sh	Sherwood number $\frac{k_m L}{D}$, analogous to Nusselt number for mass transfer
T	Temperature (Absolute) (K)
t	Time
V	Voltage

Part I
The Case for Digital Transformation of Chemical Process Industry

Introduction

1.1 The Reader of this Book

The essence of a book lies in its relevance to its audience. With this in mind, this book begins by carefully considering its target readership—professionals in technical and decision-making roles within the chemical process industry. These individuals are engaged in various aspects of chemical manufacturing and production while navigating the challenges and opportunities presented by digital transformation in this evolving landscape.

For those already acquainted with the promise and premise of digital transformation, the book's title likely serves as an immediate point of interest. The so-called "Fourth Industrial Revolution" has been reshaping the industry over the past decade, influencing multiple facets of daily operations [1]. Many organizations are actively implementing digital transformation initiatives, fostering a growing emphasis on data-driven decision-making. Process engineers, production managers, and planners are witnessing the increasing integration of automation, data acquisition, processing systems, dashboards, predictive maintenance, and various software applications into the fabric of modern production environments. Furthermore, with the advent of generative AI and the surrounding discourse, there is an emerging concern about how technical expertise will continue to remain relevant in the years ahead.

Within the industrial sector, an influx of information and considerable hype has surrounded digital transformation. Terms such as "digitalization," "Industrie 4.0," "Fourth Industrial Revolution," "Industrie 5.0," and "Smart Manufacturing" have become ubiquitous since these concepts gained traction between 2011 and 2014 [2, 3]. Corporate reports frequently reference digital transformation, integrating it into both near- and long-term strategic visions. Whether in information technology, corporate decision-making, efficiency enhancement, risk management, or sustainability initiatives, digitalization has played a pivotal role in shaping industrial progress. Alongside this movement, technical terms such as the Internet of Things (IoT), decision support systems (DSS), and decision automation (DA) have become part of the industry's common lexicon. Faced with this deluge of information, many professionals find it challenging to distinguish between reality and marketing-driven exaggeration—a sentiment shared by many in the field.

Amidst this transformation, professionals are either being ushered into an organization-wide digitalization effort or are actively involved in leading and executing such initiatives. Historically, digital transformation has often been perceived as an information technology (IT)-driven endeavor, similar to other business functions that have undergone computerization. The factory floor, once dominated by mechanical and manual processes, is now being redefined by algorithms, data systems, and advanced computing. However, within the chemical processing industry, automation and digital control have been integral to plant operations for over six decades. These advancements have significantly enhanced safety by reducing human exposure to hazardous environments. This raises important questions: What necessitates this new wave of digital transformation? What gaps exist within existing automation frameworks? What improvements can these emerging technologies truly offer? What defines a "smart plant," and does it imply that human expertise has been insufficient in ensuring efficiency and profitability? Most importantly, the pressing existential concern arises—will these digital solutions ultimately replace human roles in the process industry, leading to either obsolescence or subservience to artificial intelligence-driven systems in the manufacturing environments of the future?

For those who see their professional experiences reflected in these questions, this book aims to be a valuable resource. Its primary objective is to distill the insights gained from the authors' decade-long journey through digital transformation in the process industry. Initially inspired by the transformative promise of the Fourth Industrial Revolution, the authors embarked on numerous projects, analyzed outcomes, and accumulated a balanced perspective through both successes and failures. This book encapsulates those lessons, shaping its unique structure and narrative approach.

Written from the perspective of practitioners with backgrounds in chemical engineering and extensive experience in the process manufacturing sector, this book adopts a language distinct from other literature on digital transformation. Unlike works grounded in information technology, data science, computational science, statistics, machine learning, or artificial intelligence, this book seeks to bridge these algorithmic, mathematical, and statistical disciplines with the hands-on expertise of technical professionals in the process industry [4]. For such practitioners, this book serves as a guide to understanding the rationale and scope of digital transformation within a factory setting. It offers a structured approach to implementing digitalization in production environments, enabling professionals to monitor and measure outcomes, avoid common pitfalls, separate myths from reality, and ultimately become catalysts for the ongoing industrial revolution—whether it be the fourth, fifth, or the next iteration of transformative change in their organizations.

1.2 The Chemical Processing Industry

Chemical manufacturing is one of the most resource-intensive industries worldwide. It involves complex processes that demand substantial energy and raw materials. Petrochemical feedstocks, such as crude oil and natural gas, serve as both raw materials and energy sources. Additionally, water is used extensively throughout production, generating wastewater that may contain harmful byproducts. Emissions from various chemical processes contribute significantly to environmental pollution, raising concerns about their impact.

The industry spans a vast range of products, from mined materials to synthetic alloys, fertilizers to food, biotechnology to pharmaceuticals, petrochemicals to polymers, cosmetics to textiles, and batteries to semiconductors. As a cornerstone of modern industrial society, chemical manufacturing underpins countless sectors.

Unlike conventional manufacturing, which often evokes images of solid mechanical processing—such as metal forming, cutting, or assembly—chemical plants are characterized by an intricate network of pipes, vessels, and processing units. These facilities primarily handle liquids and gases, transferring them through pipelines, grinding solids into fine powders, storing them in tanks, and processing mixtures in large, often pressurized, high-temperature vessels. Separation techniques like filtration or evaporative removal of volatile components further refine these materials. Chemical reactions frequently accompany these processes, collectively referred to as unit operations and processes. These interconnected steps continuously transform raw materials into finished products.

In many modern industries, chemical processing is deeply integrated with mechanical manufacturing. For example, automobile production involves chemical processes such as glass tinting and exterior painting. Batteries with novel chemistries power electric vehicles, while ultrapure water treatment is essential in semiconductor fabrication. Chemical processes like chemical-mechanical polishing and circuit board etching rely on electrochemical reactions and specialized baths. Even conventional combustion-based power plants incorporate catalytic conversion to manage emissions within regulatory limits. As industrial technologies evolve, the distinction between mechanical and chemical manufacturing is increasingly blurred, forming a unified production landscape.

The petrochemical sector, closely tied to chemical production, is a major consumer of energy and raw materials. Extracting and refining fossil fuels like crude oil and natural gas are energy-intensive, significantly contributing to the industry's environmental footprint. Similarly, steel production demands vast amounts of energy and raw materials. The traditional blast furnace process, which uses iron ore, coke, and limestone, generates substantial carbon emissions, prompting a shift toward cleaner and more sustainable methods.

The food and beverage and textile industries are among the most water-intensive manufacturing sectors. Over the past few decades, significant efforts have been made to reduce their water consumption and environmental impact.

A defining characteristic of modern chemical manufacturing is its adoption of advanced technologies, including process automation, real-time monitoring, in-line analytics for quality control, and sophisticated simulations of reactive transport processes. Coupled with an expanding knowledge base in molecular chemistry, catalysis, and electrochemical reactions, these innovations have revolutionized efficiency, reduced costs, and enhanced safety. Automation minimizes human error, ensuring precision and consistent product quality, while data-driven analysis optimizes processes, predicts maintenance needs, and streamlines supply chains, making production more agile and responsive.

Sustainability has become a central focus in response to growing environmental concerns. Over the past three decades, the industry has embraced cleaner technologies and circular economy principles, reducing its ecological footprint. Initiatives such as green chemistry—which emphasizes designing processes and products with minimal environmental impact—have gained prominence. Many water-intensive sub-sectors have significantly lowered their resource consumption, prioritizing efficiency and sustainability.

Safety and regulatory compliance are also paramount. Stricter regulations and increased awareness of chemical hazards have driven companies to invest in robust safety measures and adhere to stringent guidelines. Technological advancements, such as sensor-based real-time monitoring systems, have further enhanced safety protocols, mitigating risks and ensuring secure working environments.

In the 21st century, the chemical processing industry continues to evolve, driven by technological progress, sustainability imperatives, global collaboration, and a commitment to safety. As we move forward, innovation and responsible practices will shape the industry's ongoing transformation, meeting the challenges of a rapidly changing world.

1.3 Resource Intensity of Process Industries

It is no secret that chemical manufacturing industries are often resource-intensive, placing significant demands on energy, water, and raw materials. In recent years, there has been a growing emphasis on sustainable practices within these sectors, with a focus on minimizing resource consumption and mitigating environmental impact. Governments and companies worldwide are responding to severe environmental, social and political pressures by setting ambitious net-zero targets within the next 30 years. This rapid timeline will witness not only a profound transformation of our energy landscape but also a fundamental reshaping of the entire chemical industry [5].

The chemical industry evolved in an era of cheap and abundant fossil fuel energy in the first half of the twentieth century, when the effect of greenhouse gas emissions on the climate were unknown, and the adverse effects of uninhibited fossil fuel use were not a major concern. Many of the core processes used to manufacture the industry's platform chemicals (olefins, aromatics, methanol, ammonia, etc.) relied on high temperatures and pressures, and were energy-intensive. Improvements, optimizations, and integrations were subsequently developed, with these processes progressively achieving higher efficiencies, but the key unit operations have largely remained unchanged.

1.3 Resource Intensity of Process Industries

Even though the chemical industry accounts for only approximately 7–9% of the global fossil fuel consumption, it has to seriously rethink it's feedstocks due to pressures of achieving net-zero scenarios. This necessitates negative carbon dioxide emissions, meaning that the carbon embedded in products should be of non-fossil origin. Furthermore, the oil fraction (naphtha) currently feeding the chemical industry will become economically prohibitive with the decarbonization of the transport and energy sector, resulting in a pursuit of finding other feedstocks.

These key environmental and economic drivers are underpinning the onset of a new, greener chemical industry shifting from the current use of fossil fuels as energy and feedstocks to sustainable feedstocks powered by renewable energy. New conversion processes that utilize biomass, carbon dioxide, recyclable materials, air (as a nitrogen source) and water (for hydrogen production) for the synthesis of conventional and novel platform chemicals, ultimately leading to alternative, reusable, recyclable, and biodegradable end products are being actively sought and developed.

Separation, purification and enrichment are fundamental processes in chemical manufacturing. Equally important is the development of energy-efficient separation processes. New selective and robust materials, coupled with deeper mechanistic understanding, will progressively facilitate the deployment of novel, energy-efficient separation techniques, including adsorption, absorption, and membrane separation. These processes will help reduce dependence of the chemical manufacturing industry on traditional energy-intensive thermal separations like distillation, evaporation and drying that currently dominate the industry and consume approximately 10–15% of the global energy.

The focus on energy efficiency, reduction in use of fossil fuel based feedstocks, lower GHG emissions and net-zero targets are already ushering in significant paradigm shifts in the chemical process industry. For example, the integration of power generation and water recycle are becomeing commonplace in the upcoming chemical process plants, blurring the limits between the chemical, energy, and utility industries. Unlike the existing model of continuous fossil fuel based energy supply, new processes will need to adapt to the seasonal and intermittent fluctuations in renewable energy supply, necessitating innovative solutions beyond conventional buffering methods (for instance, using batteries or other traditional energy storage methods). Such scenarios will necessitate flexible operation where the processes are capable of swiftly adjusting production capacity, marking a departure from the conventional steady-state operation in many chemical processing plants. Achieving this flexibility in production requires exploration of novel control strategies, robust equipment, processes amenable to cyclic operating conditions, and smart scheduling protocols. The diffuse nature of renewable energy and its demand for large surface areas for its production may also influence the configuration of the future green chemical industry, leading to smaller processes and more distributed, local manufacturing. Similarly, new heat integration, transfer, and storage approaches will be needed that venture beyond the current reliance on steam.

1.4 Circular Economy Practices

The chemical manufacturing industry is at the forefront of adopting circular economy principles to address environmental challenges while ensuring economic viability. At its core, the circular economy emphasizes conservation, efficient resource utilization, and sustainability, with an ultimate goal of achieving net zero emissions. These principles are increasingly shaping the industry's transformation, fostering innovation and inter-sector collaboration. In the chemical and petrochemical sectors, for example, the development of closed-loop systems helps recover and reuse certain byproducts, minimizing the need for virgin raw materials. Steel producers are exploring innovative methods such as electric arc furnaces and hydrogen-based steelmaking to reduce carbon emissions and resource use.

Conservation lies at the heart of the circular economy. In chemical manufacturing, this translates to optimizing processes to minimize waste and energy consumption. Advanced technologies, such as process intensification and real-time monitoring, allow manufacturers to reduce raw material usage and improve yields. Recycling and reusing materials, like solvents and catalysts, further exemplify conservation efforts. Additionally, employing renewable feedstocks, such as biomass, reduces dependency on finite resources like fossil fuels.

The industry is increasingly embracing resource-sharing practices, where byproducts from one process become valuable inputs for another. This cross-sector utilization maximizes resource efficiency and minimizes waste. For example, CO_2 captured from chemical plants is used in the food and beverage industry or as a precursor for producing synthetic fuels and polymers. Similarly, waste heat from chemical processes is harnessed to generate energy for nearby industries or communities. Produced water, a major by-product from oil exploration activities, is being treated and reused as a resource for other industrial water needs, agriculture, irrigation, or dust suppression. Concentrated brine generated from seawater desalination is being increasingly seen as a source of minerals such as magnesium, potassium, and lithium. Such synergies create a network of interdependent sectors that operate in a closed-loop system.

Sustainability in chemical manufacturing is achieved through a combination of green chemistry, eco-friendly designs, and energy-efficient practices. Green chemistry principles guide the development of safer, non-toxic, and biodegradable chemicals, minimizing environmental and health risks. Sustainable production methods, such as enzymatic catalysis and electrochemical synthesis, reduce reliance on high-energy processes. Moreover, adopting renewable energy sources, such as solar and wind, for chemical plants significantly lowers the industry's carbon footprint.

The ultimate aspiration of the chemical manufacturing industry within the circular economy is to achieve net zero emissions, if not during manufacturing of a product, then at least over the life cycle of the product. This involves not only reducing direct emissions but also offsetting residual emissions through carbon capture, utilization, and storage (CCUS) technologies. Advanced material recovery systems, waste-to-energy technologies, and the development of carbon-neutral products are crucial steps toward this goal. Collaborative efforts with governments and other industries further strengthen the drive toward a sustainable and climate-resilient future. By integrating circular economy principles, the chemical manufacturing industry is paving the way for a sustainable, resource-efficient, and net-zero future, fostering resilience in a rapidly changing world.

1.5 Digital Transformation in the Chemical Industry

The chemical manufacturing industry has long been a pioneer in adopting advanced technologies to optimize operations and enhance productivity. Its journey toward digital transformation began decades ago with the introduction of process automation and control systems. The development of Distributed Control Systems (DCS) in the 1970s and 1980s marked a significant milestone, enabling centralized monitoring and control of complex chemical processes. These early digital control systems laid the groundwork for modern-day digitalization, embodying the industry's forward-thinking approach to technology integration.

Digital transformation or digitalization entails creating a network of cyber physical systems (CPS) integrated to a factory or a manufacturing plant that can enhance the manufacturing operation into a "smart operation". The concept involves connecting the physical plant to a digital counterpart (digital realm) by collecting information from the plant utilizing sensors or monitoring devices. These sensors digitize the information and transmit the data through a variety of communications mechanisms (wireless, local area network, cellular transmission, etc.) to the digital realm. These network of sensors and connected devices is collectively referred to as the internet of things (IoT). The data from different sections of the plant can be collected into a data repository, which acts as a consolidated single source of truth. This database can provide information about the entire plant and production operation allowing transparency and interoperability between different sections of the plant. In conjunction with other types of databases of a business, the plant data can be processed to provide information about the production, plant efficiency, maintenance, supply chain, and inventory. Suitably processing the data can lead to insights about the various working components of the manufacturing process simultaneously. This interoperability can provide insights that can assist the plant operators to manage the operations of the plant more efficiently. This aspect is referred to as decision support. Finally, the physical and digital infrastructures can be connected in a loop, whereby the digital realm insights about the process can be utilized to autonomously (without human intervention) operate and control the plant. This feature is referred to as decision automation. The overall result of closing this cyber physical loop is an enhanced level of automation and adaptability of the manufacturing system.

The essence of digital transformation was first embodied in the Industrie 4.0 concept originally proposed in Germany around 2011. Industrie 4.0 embodies the interactive cyber physical system design principles that allow achieving these four levels of sophistication in industrial manufacturing operations, namely, transparency, interoperability, decision support, and decision automation [2, 3, 6].

Digital transformation in the chemical manufacturing industry entails comprehensive integration of digital technologies across all aspects of production and operations. In the context of smart plants, this transformation aligns with the principles of Industrie 4.0, which emphasizes automation, interconnectivity, and real-time data analytics. Smart plants leverage technologies such as the Internet of Things (IoT), Artificial Intelligence (AI), and cloud computing to create self-optimizing production environments. Industrie 5.0 builds on this foundation by promoting human-centric innovation [7, 8], where advanced technologies complement human decision-making, fostering collaboration between operators and machines. The result is an industry that is not only efficient and data-driven but also adaptable and resilient, capable of meeting evolving demands and sustainability goals.

The chemical industry's historical leadership in implementing digital control systems (DCS) provides a strong foundation for embracing digital transformation. Many chemical plants are now integrating traditional Operational Technology (OT) layers, such as process control and manufacturing execution systems (MES), with modern Information Technology (IT) platforms. This convergence enables seamless data flow across the value chain, facilitating predictive maintenance, process and energy optimization, and advanced process modeling.

Digital transformation of a chemical processing plant requires an assessment of the pre-digitalization state of automation of the plant. In this regard, most sectors of chemical processing involve quite advanced digital control systems (DCS). Traditional chemical processing plants deploy advanced process control methods to meet desired quality of the end products, ensure safe operation, adhere to environmental regulations, and attain economic viability. Traditionally, proportional- integral- derivative- (PID) control and model predictive control (MPC) have been the cornerstones of advanced digital control of chemical processes [9, 10]. PID control implements a feedback or feedforward loop to adjust a manipulated variable based on the drift of a measured variable from its target set-point. The MPC approach, a constrained optimization-based control method, utilizes time-series data-based linear dynamic models, and is used to compute control actions to maintain optimal process operation while accounting for process and control actuator constraints. However, several limitations exist in these PID and MPC frameworks. First, PID loops are primarily designed for single parameter controls, and inherently assume independence of causalities and linear superposition principles when considering influence of multiple factors on a process. MPC considers local linear excursions of the variables from a desired control set-point. Many chemical processes are inherently nonlinear. While these methods are staples for control of individual process components or systems, their incorporation in an integrated train of several connected unit operations and processes can be problematic. This poses a serious obstacle toward achieving the target goals of plant-wide digitalization, such as interoperability and transparency between multiple unit operations.

1.5 Digital Transformation in the Chemical Industry

Digital twins—physics based virtual replicas of processes—are increasingly being used to simulate plant operations, allowing operators to optimize performance and troubleshoot issues without disrupting production [11, 12]. Similarly, artificial intelligence (AI) driven analytics help predict equipment failures, enhance process efficiency, and ensure product quality. By adopting these practices, the chemical sector is unlocking new levels of productivity and innovation. While first-principles-based modeling via digital twins provides a direct way for nonlinear dynamic process model development, it may be cumbersome and difficult to implement in complex industrial processes that may also not be well understood from a fundamental physicochemical standpoint. As an alternative, data-driven machine learning (ML) modeling, with emphasis on recurrent neural network configurations trained using time-series data and ensemble learning schemes, provides a new paradigm for nonlinear dynamic model development that can be used in the implementation of nonlinear MPC systems [13, 14]. Despite this progress, dataset construction and pre-processing, network construction using process-specific knowledge, network error quantification and real-time update are important yet unresolved issues that are crucial in the field of ML-based MPC. While nascent research efforts have already been dedicated to these problems, a significant set of challenges remains to be addressed [15].

In addition to revealing nonlinear dynamic process relationships from data, ML tools can address classification problems such as the ones arising in the context of process operational safety as well as fault diagnostics and prognostics, thus providing a broad array of topics where ML can make an impact [16]. Process operational safety is possibly the most crucial reason to implement controllers in a chemical process. For instance, if the temperature or pressure within a reactor exceeds its material limits and leads to an explosion, the results can be catastrophic in terms of lives, the environment and capital. To incorporate safety directly into the controller design, advanced formulations of MPC (such as a control-Lyapunov-barrier-function-based MPC) have been proposed [17], which allow the delineation of specific, unsafe regions of operation to be avoided by the controller, the knowledge of which may be obtained from operators and engineers. While such methods typically require *a priori* knowledge of a well-defined region in the operating space, ML-based classification methods can identify such unsafe regions from process operational data, regardless of the complexity of the unsafe region, and should be explored further.

In addition to these long-standing issues, the rapid proliferation of networked sensor and actuation devices in the context of the Industrie 4.0 has made cybersecurity a growing concern that requires novel approaches to be handled.

In the current era of big data, process safety is also strongly linked to the control system security as control systems that utilize an array of networked sensors and actuators are vulnerable to cyberattacks. Intelligent cyberattacks that are aimed at deteriorating closed-loop performance without destabilizing the process can require advanced defensive methods such as ML-based detectors that employ neural networks to detect and classify these attacks. To further secure control systems, the networked communications within the system may also be encrypted to establish cyber-secure networked communications. However, due to the high complexity and computational cost of MPC for large-scale processes, decentralized and distributed ML detectors and MPC using encrypted networked communications may be needed, requiring further research and academic focus in this direction [18].

Digital transformation is instrumental in driving the chemical industry toward circular economy and net zero goals. Smart plants equipped with real-time monitoring and advanced analytics can minimize waste, optimize energy usage, and enable the recovery and recycling of materials. Technologies such as carbon capture, utilization, and storage (CCUS) are integrated into digital platforms, allowing for efficient tracking and management of emissions.

Furthermore, the interconnectivity facilitated by digital transformation supports cross-sector collaboration, where waste streams from one industry can be repurposed as inputs for another. By embracing circular economy principles, chemical manufacturers can reduce their environmental footprint while maintaining economic viability.

The ongoing digital transformation in the chemical manufacturing industry represents a paradigm shift toward sustainable, efficient, and interconnected production systems. By building on its legacy of innovation in process automation and control, the industry is well-positioned to lead the transition to smart factories. This transformation not only enhances operational efficiency but also aligns with global efforts to achieve a circular economy and net zero emissions, ensuring a sustainable future for the industry and the planet.

In conclusion, the era of the development and implementation of data-driven tools in process control, safety and operations has begun, and it promises to dramatically improve our ability to operate and control industrial chemical processes.

References

1. Ray Y. Zhong, Xun Xu, Eberhard Klotz, and Stephen T. Newman. Intelligent manufacturing in the context of industry 4.0: A review. *Engineering*, 3(5):616–630, 2017.
2. Li Da Xu, Eric L. Xu, and Ling Li. Industry 4.0: State of the art and future trends. *International Journal of Production Research*, 56(8):2941–2962, 2018.
3. Robert Lawrence Wichmann, Boris Eisenbart, and Kilian Gericke. The direction of industry: A literature review on industry 4.0. *Proceedings of the Design Society: International Conference on Engineering Design*, 1(1):2129–2138, 2019.
4. Pauliina Rikala, Greta Braun, Miitta Järvinen, Johan Stahre, and Raija Hämäläinen. Understanding and measuring skill gaps in industry 4.0 — a review. *Technological Forecasting and Social Change*, 201:123206, 2024.
5. Laura Torrente-Murciano, Jennifer B. Dunn, Panagiotis D. Christofides, Jay D. Keasling, Sharon C. Glotzer, Sang Yup Lee, Kevin M. Van Geem, Jean Tom, and Gaohong He. The forefront of chemical engineering research. *Nature Chemical Engineering*, 1(1):18–27, 2024.
6. H. Lasi, P. Fettke, H.-G. Kemper, T. Feld, and M. Hoffmann. Industry 4.0. *Business & Information Systems Engineering*, 6:239–242, 2014.
7. Amr Adel. Future of industry 5.0 in society: human-centric solutions, challenges and prospective research areas. *Journal of Cloud Computing*, 11(1):40, 2022.

References

8. Praveen Kumar Reddy Maddikunta, Quoc-Viet Pham, Prabadevi B, N Deepa, Kapal Dev, Thippa Reddy Gadekallu, Rukhsana Ruby, and Madhusanka Liyanage. Industry 5.0: A survey on enabling technologies and potential applications. *Journal of Industrial Information Integration*, 26:100257, 2022.
9. B. Wayne Bequette. *Process Control: Modeling, Design, and Simulation*. Prentice Hall, Upper Saddle River, NJ, USA, first edition, 2002.
10. Dale Seborg, Thomas Berry, and G. Victor Reklaitis. *Process Dynamics and Control*. Wiley Series in Chemical Engineering. John Wiley & Sons, Hoboken, NJ, USA, third edition, 2011.
11. Mark Asch, Olivier P. Le Maître, and Jérémie Mary. *A Toolbox for Digital Twins: From Model-Based to Data-Driven*. SIAM, 2022.
12. Michael Klotz, Kai Wagemann, Jürgen Weber, André Bühler, and Jürgen Beyerer. *Generation and Update of a Digital Twin in a Process Plant*. Springer, 2023.
13. Niket Sharma and Y. A. Liu. A hybrid science-guided machine learning approach for modeling and optimizing chemical processes. *arXiv preprint* arXiv:2112.01475, 2021.
14. Andreas Himmel, Janine Matschek, Rudolph Kok, Bruno Morabito, Hoang Hai Nguyen, and Rolf Findeisen. Machine learning for process control of (bio)chemical processes. *arXiv preprint* arXiv:2301.06073, 2023.
15. Malte Esders, Gimmy Alex Fernandez Ramirez, Michael Gastegger, and Satya Swarup Samal. Scaling up machine learning-based chemical plant simulation: A method for fine-tuning a model to induce stable fixed points. *arXiv preprint* arXiv:2307.13621, 2023.
16. Marcin Pietrasik, Anna Wilbik, and Paul Grefen. The enabling technologies for digitalization in the chemical process industry. *Digital Chemical Engineering*, 12:100161, 2024.
17. Zhe Wu, Fahad Albalawi, Zhihao Zhang, Junfeng Zhang, Helen Durand, and Panagiotis D. Christofides. Control lyapunov-barrier function-based model predictive control of nonlinear systems. *Automatica*, 109:108508, 2019.
18. Prodromos Daoutidis, Larry Megan, and Wentao Tang. The future of control of process systems. *Computers & Chemical Engineering*, 178:108365, 2023.

The Business Case for Digital Transformation

2.1 Introduction

Digital transformation in any industry is driven by a fundamental objective: to enhance efficiency, productivity, and agility, ultimately creating a smarter, more adaptive system. However, for these efforts to be viable, they must make business sense—justifying the investment and delivering a measurable return on the investment. When considering the digital transformation within a specific sector of the chemical industry, of a chemical processing plant, or within a component of the plant, one must evaluate how these advancements align with the industry's vision and evolving demands. This requires a keen understanding of how the chemical process industry (CPI) will evolve over the coming decades, the economic and geopolitical landscape that will shape its trajectory, and the key challenges that must be addressed during the life cycle of the project. Developing a forward-looking vision for the industry and strategically integrating digitalization efforts to support that vision is not just beneficial—it is essential for any meaningful transformation. In this chapter, we briefly outline the pertinent guiding paradigms for the CPI, and how digitalization efforts need to conform to these overarching paradigms of the industry.

2.2 Chemical Process Industry Paradigms

2.2.1 Sustainability

The chemical processing industry, long associated with innovation and progress, is now at the forefront of a transformative journey toward sustainability. As the global community increasingly recognizes the environmental impact of industrial activities, chemical manufacturers are embracing sustainable practices to mitigate their footprint. This shift towards sustainability in the chemical processing industry is not only an ethical imperative but also a strategic response to evolving market demands and regulatory pressures.

2.2.2 Reducing Carbon Footprints

One of the central tenets of sustainability in the chemical processing industry involves the reduction of carbon emissions. Companies are adopting cleaner and more energy-efficient technologies, transitioning to renewable energy sources, and implementing carbon capture and storage solutions. These initiatives not only contribute to global efforts to combat climate change but also enhance operational efficiency and long-term cost-effectiveness.

2.2.3 Circular Economy and Resource Efficiency

Embracing a circular economy model is another key facet of sustainable practices in chemical processing. This approach emphasizes the reduction, reuse, and recycling of materials to minimize waste generation. Manufacturers are increasingly investing in closed-loop systems, where byproducts and waste from one process become feedstocks for another. This not only conserves resources but also reduces the environmental impact of raw material extraction and disposal.

2.2.4 Green Chemistry and Safer Alternatives

The adoption of green chemistry principles is driving the development of sustainable processes and products. Manufacturers are exploring safer alternatives to traditional chemicals, opting for substances that are less hazardous to human health and the environment. Innovations in green chemistry aim to design processes that generate minimal waste, use renewable feedstocks, and prioritize energy efficiency.

2.2.5 Water and Energy Conservation

Sustainable practices in the chemical processing industry also focus on water and energy conservation. Companies are implementing water recycling and purification systems, minimizing water usage in production processes. Furthermore, energy-efficient technologies, such as advanced heat exchangers and process optimization, are being adopted to reduce energy consumption and improve overall efficiency.

2.2 Chemical Process Industry Paradigms

2.2.6 Transparency and Responsible Sourcing

Transparency in the supply chain and responsible sourcing of raw materials are integral components of sustainable practices. Companies are increasingly scrutinizing their supply chains to ensure that raw materials are ethically sourced, minimizing the environmental and social impact. This approach not only aligns with ethical standards but also builds trust with environmentally conscious consumers and stakeholders.

2.2.7 Regulatory Compliance and Industry Collaboration

Governments and regulatory bodies are playing a pivotal role in steering the chemical processing industry toward sustainability. Stringent environmental regulations and emission standards are compelling companies to adopt greener practices. Moreover, industry collaboration through initiatives, such as Responsible Care, fosters knowledge sharing and the development of best practices for sustainability across the sector.

2.2.8 Challenges and Future Outlook

The industry paradigms outlined in this section are much evolved and refined, probably completely alien to the CPI business paradigms of fifty, thirty, or even twenty years in the past. It shows how industrialization has evolved and modernized along with humanity in the past five decades. In some sense, it appears that profitability as a goal is superseded in the present day by the need to have a social license to conduct business responsibly. This is true for the entire industrial world of the 21st century, not just the CPI. This also is universal practice across the globe, and not just limited to certain regions or geopolitical landscapes.

While progress is evident, the chemical processing industry still faces challenges in achieving comprehensive sustainability. Balancing economic viability with environmental responsibility, navigating the complexities of global supply chains, and overcoming technical barriers to sustainable innovation are ongoing challenges. However, these challenges present opportunities for continued research, innovation, and collaboration to create a more sustainable and resilient industry.

By prioritizing environmental stewardship, embracing circular economy principles, and fostering a culture of innovation, chemical manufacturers are proving that sustainable practices not only protect the planet but also ensure long-term success in an evolving marketplace. As sustainability becomes an integral part of the industry's operating paradigm, the chemical processing sector is poised to lead the way towards a more sustainable and responsible future.

2.3 Human Aspects

Businesses are conceived, owned, and operated by humans, ultimately providing value to both individuals and society. As a result, there exists a complex and evolving interaction between businesses and humankind. Over the past few decades, this interaction has become increasingly global, ubiquitous, and dynamic. The chemical industry, in particular, has experienced this transformation more profoundly than most.

A few decades ago, business production environments were largely hidden from customers. Today, they are more transparent than ever, seamlessly connected to consumers through advancements in information technology. Consumers can now access real-time insights into business operations and practices, while their preferences evolve dynamically—tracked almost instantaneously by marketing teams. This creates a continuously accelerating feedback loop between businesses and consumers, driving rapid cycles of product updates, innovation, and customization.

For instance, consider how customers today can personalize a product online, triggering an immediate manufacturing response. Modern manufacturing has become agile and highly responsive, with product customization becoming the norm in many consumer markets. Traditional manufacturing, shipping, and warehousing models have given way to just-in-time production, on-demand customization, and globally optimized logistics, fundamentally reshaping the industrial landscape.

These manufacturing systems are what we refer to in this book as human-in-the-loop (HITL) systems. In the above, we have described primarily the consumers as humans in the loop. However, humans are deeply ingrained in other aspects of the manufacturing systems, namely, business operations, and the manufacturing environments.

Within the manufacturing environment, human-in-the-Loop (HITL) systems represent a dynamic fusion of human intelligence and machine capabilities, fostering a collaborative environment where humans and machines work together to achieve complex tasks. These systems are designed to capitalize on the strengths of both entities, leveraging human intuition and adaptability alongside machine precision and efficiency. Navigating the nuances of human-machine interactions in HITL systems involves understanding the delicate balance between automation and human intervention, ensuring optimal performance across various domains.

2.3.1 Human-in-the-Loop Systems

Human-in-the-Loop systems are characterized by their ability to seamlessly integrate human decision-making with automated processes. These systems recognize that while machines excel at repetitive tasks and data processing, human cognition brings unique qualities such as creativity, context understanding, and emotional intelligence. The synergy between humans and machines becomes particularly crucial in scenarios where complex decision-making, adaptability, and ethical considerations are paramount.

One of the critical nuances in HITL systems is finding the right balance between automation and human control. The degree of autonomy granted to machines depends on the nature of the task and the level of human expertise required. For example, in autonomous vehicles, humans may intervene in critical situations, striking a balance between automated navigation and human decision-making.

2.3.2 Adaptive Learning and Continuous Improvement

HITL systems incorporate adaptive learning mechanisms to enhance their performance over time. Humans provide crucial feedback and corrections to the system, allowing it to refine its algorithms and decision-making processes. This continuous loop of feedback and improvement is essential for staying agile in dynamic environments and adapting to evolving challenges.

2.3.3 Enhancing User Experience and Trust

Ensuring a positive user experience is another nuanced aspect of human-machine interactions in HITL systems. User interfaces must be intuitive, providing humans with clear insights into the system's operations. Building trust between humans and machines is crucial, and transparent communication about system capabilities, limitations, and decision-making processes fosters confidence and facilitates collaboration.

2.3.4 Domain-Specific Considerations

The nuances of human-machine interactions in HITL systems vary across domains. In healthcare, for instance, HITL systems may assist in medical diagnostics, with human experts providing insights and interpretations that machines may not discern alone. In cybersecurity, human analysts may collaborate with automated threat detection systems to identify and respond to sophisticated cyber threats.

2.3.5 Ethical and Social Implications

As HITL systems become increasingly prevalent, ethical considerations surrounding privacy, bias, and accountability come to the forefront. Human intervention is essential for addressing ethical dilemmas and ensuring that machines align with societal values. The nuanced collaboration between humans and machines should be guided by ethical frameworks to navigate potential pitfalls and ensure responsible use of technology.

The nuances of human-machine interactions in Human-in-the-Loop systems exemplify the intricate dance between human expertise and machine efficiency. As these systems continue to evolve, understanding and addressing the complexities surrounding autonomy, user experience, adaptive learning, and ethical considerations are paramount. By fostering a harmonious collaboration between humans and machines, HITL systems promise to unlock new realms of possibilities, offering solutions that leverage the unique strengths of both entities.

2.3.6 Role of Digital Transformation in HITL Systems

Digital transformation is reshaping the landscape of the chemical process industry, revolutionizing human-machine interactions and fostering unprecedented levels of efficiency, safety, and innovation. Across various sectors such as petroleum refining, food and beverage processing, chemical processing, and mining and mineral processing, the integration of digital technologies is transforming how humans interact with machines, resulting in smarter processes and more informed decision-making.

2.3.7 Challenges and Opportunities

While the benefits of human-machine interaction in the chemical process industry are evident, challenges such as cybersecurity and workforce adaptation must be addressed. Companies are investing in robust cybersecurity measures to protect sensitive data and critical infrastructure. Workforce training programs are crucial to ensure that personnel can effectively operate and leverage the capabilities of digital technologies.

As the chemical process industry embraces digital transformation, human-machine interactions are evolving to new heights of sophistication and efficiency. From augmented reality in training to advanced process control systems, these technologies are shaping a future where humans and machines work seamlessly together. The integration of digital transformation technologies not only enhances safety and efficiency but also positions the industry at the forefront of innovation, ensuring its competitiveness in a rapidly changing global landscape.

2.4 Process Industry and Information Technology

The rapid evolution of information technology (IT) has profoundly reshaped the business landscape, particularly in the process industry. Once reliant on rigid, labor-intensive operations, businesses now operate in a highly digitized and interconnected world where data,

2.4 Process Industry and Information Technology

automation, and real-time decision-making drive efficiency and competitiveness. IT has revolutionized how businesses function, enabling smarter operations, global connectivity, and enhanced adaptability in a dynamic marketplace.

In the process industry—encompassing chemical manufacturing, oil and gas, pharmaceuticals, and food processing—IT has transformed every facet of operations. Advanced data analytics, artificial intelligence, and real-time monitoring systems have optimized production, reduced waste, and improved safety. Process automation in the factory environment, enabled by distributed control systems (DCS), programmable logic controllers (PLCs), and supervisory control and data acquisition (SCADA) systems, has minimized human intervention while maximizing efficiency. Collectively referred to as operating technology (OT), the DCS and its elements were primarily confined within the factory environment, and disconnected from the business operations. With the integration of IT, and associated technologies such as enterprise resource planning (ERP) and manufacturing execution systems (MES) created an amalgamation of OT with IT, ensuring seamless coordination between production, supply chain, and business functions, fostering a more responsive and agile industry.

IT is the driving force behind the digitalization of the manufacturing sector. The rise of digital twins—virtual replicas of physical processes—allows companies to simulate, predict, and optimize operations with unprecedented precision. The Industrial Internet of Things (IIoT) connects equipment, sensors, and cloud-based platforms, enabling real-time data-driven decision-making. Predictive maintenance, powered by machine learning algorithms, reduces downtime and enhances asset longevity. These advancements not only increase productivity but also enable greater sustainability by minimizing resource consumption and emissions.

Ultimately, information technology is not just a tool but the backbone of modern industrial transformation. Businesses that embrace IT-driven digitalization gain a significant competitive edge, ensuring resilience, innovation, and long-term success in an increasingly complex and fast-paced global market. As industries continue to evolve, IT will remain the cornerstone of progress, driving the next wave of intelligent, autonomous, and sustainable manufacturing.

In the ever-evolving landscape of the process industry, the integration of Information Technology (IT) has emerged as a transformative force, reshaping traditional operations and propelling the sector toward new frontiers of efficiency and innovation. The marriage of process industry and IT has given rise to a dynamic synergy, empowering companies to optimize operations, improve safety, and foster innovation.

2.4.1 Enhancing Efficiency Through Automation

One of the key contributions of IT to the process industry is the automation of various processes. Automation systems, powered by advanced control algorithms and sensors, facilitate real-time monitoring and control of industrial processes. From manufacturing plants to refineries, the automation of routine tasks not only improves efficiency but also minimizes errors, reduces downtime, and ensures consistent product quality. The seamless integration of IT solutions enables companies to achieve higher levels of precision, accuracy, and reliability in their operations.

2.4.2 Data Analytics and Predictive Maintenance

The process industry is increasingly leveraging the power of data analytics to extract meaningful insights from vast amounts of operational data. Predictive maintenance, a key application of data analytics, allows companies to anticipate equipment failures and schedule maintenance activities proactively. This not only reduces unplanned downtime but also extends the lifespan of critical assets, optimizing the overall cost of maintenance. Data-driven decision-making is becoming a cornerstone of strategic planning in the process industry, enabling companies to identify trends, optimize processes, and respond quickly to changing market conditions.

2.4.3 Integration of Industrial Internet of Things (IIoT)

The Industrial Internet of Things (IIoT) has become a game-changer in the process industry, connecting devices and systems to create a network of intelligent and interconnected assets. Sensors embedded in equipment collect real-time data, providing a comprehensive view of industrial processes. This connectivity allows for remote monitoring, predictive analytics, and the optimization of resource utilization. From smart sensors to connected machinery, the IIoT is fostering a new era of efficiency, flexibility, and responsiveness in the process industry.

2.4.4 Cybersecurity Challenges and Solutions

While the benefits of IT integration in the process industry are undeniable, it also brings forth new challenges, particularly in terms of cybersecurity. With the increased connectivity of industrial systems, the risk of cyber threats becomes a critical concern. Companies in the process industry are investing in robust cybersecurity measures to safeguard sensitive data, protect critical infrastructure, and ensure the uninterrupted flow of operations. The convergence of IT and operational technology (OT) requires a comprehensive and proactive approach to cybersecurity to mitigate potential risks.

The symbiotic relationship between the process industry and information technology is reshaping the sector's landscape, ushering in an era of heightened efficiency, innovation, and competitiveness. As companies continue to embrace digital transformation, the integration of IT solutions will remain a cornerstone for driving operational excellence, ensuring sustainability, and navigating the complexities of the modern industrial landscape. The process industry is poised for continued evolution, leveraging the power of information technology to unlock new possibilities and maintain a competitive edge in a rapidly changing global marketplace.

2.5 Digital Transformation Scenarios in Selected Sectors

2.5.1 Petroleum Refining and Petrochemicals Processing

Petroleum refining and petrochemicals businesses are integral components of the energy and chemical sectors, focusing on the processing of crude oil into refined products and the transformation of hydrocarbons into diverse chemical compounds. In petroleum refining, crude oil undergoes distillation and various refining processes to produce essential fuels like gasoline, diesel, and jet fuel. Concurrently, the petrochemical industry utilizes byproducts from refining to manufacture a broad spectrum of chemicals, such as plastics, fertilizers, and synthetic materials. Both sectors require sophisticated infrastructure, advanced technologies, and stringent safety measures. The petroleum refining and petrochemicals businesses collectively contribute significantly to global energy needs and the production of essential materials for various industries.

In the complex landscape of the petroleum refining and petrochemicals industry, digital transformation has emerged as a transformative force, revolutionizing traditional processes and opening new avenues for efficiency and sustainability. This paradigm shift is fundamentally altering how businesses operate, optimize production, and respond to market demands.

One of the primary benefits of digital transformation in this sector lies in process optimization. Advanced process control systems, powered by artificial intelligence and machine learning algorithms, enable real-time monitoring and adjustment of refining and petrochemical processes. This results in enhanced operational efficiency, reduced downtime, and improved overall plant performance. Predictive maintenance, enabled by sensors and data analytics, ensures that equipment issues are identified and addressed before they escalate, minimizing disruptions and maximizing asset utilization.

Furthermore, digital technologies play a crucial role in improving supply chain management within the petroleum and petrochemical industries. Advanced analytics and digital platforms facilitate better coordination between suppliers, manufacturers, and distributors. This streamlined communication enhances inventory management, reduces lead times, and improves overall logistics efficiency, ensuring a smoother flow of raw materials and finished products throughout the supply chain.

Digital transformation also contributes significantly to safety and environmental sustainability. Advanced monitoring systems, IoT sensors, and predictive analytics help identify potential safety hazards and environmental risks. This proactive approach allows companies to implement preventive measures, reducing the likelihood of accidents and minimizing the environmental impact of their operations. Additionally, digital technologies support the development and implementation of sustainable practices, such as energy-efficient processes and the utilization of renewable energy sources, aligning with global efforts to reduce carbon footprints.

The integration of digital twins—a virtual representation of physical assets or processes—further enhances decision-making processes in the petroleum refining and petrochemicals sector. Digital twins allow for simulations and scenario analysis, enabling companies to optimize production processes, test new strategies, and evaluate the impact of potential changes without disrupting actual operations.

Digital transformation is a game-changer for the petroleum refining and petrochemicals industry. By embracing advanced technologies, companies can unlock new levels of operational efficiency, safety, and sustainability. As the industry navigates the complexities of a rapidly changing global landscape, digital transformation becomes not only a strategic advantage but a necessity for those looking to thrive in a competitive and environmentally conscious future.

2.5.2 Food and Beverage Industry

The food and beverage industry, a cornerstone of global commerce, encompasses a vast array of businesses involved in the production, processing, and distribution of consumable goods. From agricultural cultivation and food processing to restaurant services and retail, the sector operates at multiple levels of the supply chain. This dynamic industry caters to diverse consumer preferences, constantly innovating to meet evolving tastes and dietary trends. Successful businesses within the food and beverage sector not only deliver high-quality products but also navigate complex logistics, stringent safety standards, and changing market dynamics to ensure sustained success in this competitive market. The food and beverage sector, a cornerstone of global commerce, has undergone a profound transformation through digital technologies, revolutionizing various subcategories within the industry. From production and distribution to consumer engagement, the impact of digital transformation is palpable, enhancing efficiency, sustainability, and customer experiences.

In the realm of agricultural production, precision farming technologies are optimizing crop management through the use of sensors, drones, and data analytics. These tools provide real-time insights into soil health, crop conditions, and weather patterns, allowing farmers to make informed decisions, reduce resource usage, and enhance yields. This digital revolution is not only increasing agricultural efficiency but also promoting sustainable practices.

In food processing, automation and smart manufacturing are streamlining production lines. Robotics and artificial intelligence are employed to improve accuracy, consistency, and speed in tasks such as sorting, packaging, and quality control. This not only enhances production efficiency but also ensures the delivery of high-quality products to consumers.

2.5 Digital Transformation Scenarios in Selected Sectors

Supply chain management within the food and beverage industry has benefited immensely from digital technologies. Blockchain, for instance, is being utilized to create transparent and traceable supply chains. From farm to fork, consumers can now access detailed information about the origin, processing, and transportation of their food, fostering trust and ensuring the safety of products.

Retail and hospitality have experienced a digital renaissance through the adoption of point-of-sale (POS) systems, mobile ordering apps, and personalized customer engagement platforms. These technologies enhance the overall customer experience, providing convenience, personalization, and efficient service. Additionally, data analytics enable businesses to understand consumer preferences, optimize inventory management, and design targeted marketing strategies.

The restaurant industry, in particular, has witnessed a digital revolution with the rise of online food delivery platforms and digital menus. These platforms leverage algorithms to streamline order processing, optimize delivery routes, and enhance customer satisfaction. Furthermore, digital marketing and social media play a crucial role in promoting restaurants and engaging with a broader audience.

In conclusion, digital transformation has become a catalyst for innovation and efficiency across various subcategories of the food and beverage sector. From farm to consumer, the integration of digital technologies is fostering sustainability, improving operational processes, and elevating the overall customer experience. As the industry continues to embrace the possibilities offered by digital transformation, it is poised for continued growth and adaptation to the evolving needs of consumers and the broader market.

2.5.3 Speciality Chemical and Biochemicals Processing

The specialty chemicals and biochemicals industry represents a dynamic sector within the chemical processing landscape, emphasizing the production of unique and high-value chemicals for specific applications. Specialized in the creation of additives, catalysts, polymers, and performance-enhancing compounds, this industry caters to diverse sectors including pharmaceuticals, agriculture, electronics, and manufacturing. The focus extends beyond mass production to precision and customization, addressing specific needs in various applications. In parallel, the biochemicals sector harnesses biological processes to produce chemicals from renewable resources, aligning with sustainable practices. The businesses in these industries operate at the forefront of innovation, constantly developing novel compounds to meet evolving market demands and environmental considerations.

The chemical processing industry is witnessing a profound shift in human-machine interactions through digital transformation. Process automation and control systems enable operators to oversee and adjust chemical processes with precision, reducing the risk of errors and enhancing efficiency. Virtual reality (VR) simulations assist in the training of personnel, allowing them to familiarize themselves with complex equipment and emergency scenarios in a safe, virtual environment. Advanced analytics further empower decision-makers with insights for optimizing processes and resource utilization.

2.5.4 Mining and Mineral Processing

The mining and mineral processing industry is a cornerstone of global resource extraction, encompassing activities from exploration and extraction to refining and distribution. Mining operations involve locating and extracting valuable minerals and metals from the earth, while mineral processing refines these raw materials into marketable products. This industry is characterized by complex supply chains, advanced technologies, and stringent environmental regulations. It plays a pivotal role in supplying essential materials for various sectors, including construction, manufacturing, and energy production, shaping the foundation of many economies worldwide.

In the mining and mineral processing industry, digital transformation is enhancing safety and efficiency in resource extraction. Autonomous vehicles and drones equipped with sensors and cameras enable remote monitoring of mining operations, reducing the need for personnel in hazardous environments. Wearable devices equipped with real-time tracking systems enhance worker safety, providing alerts for potential dangers. Data analytics algorithms optimize mineral processing, improving yields and reducing waste.

2.5.5 Pharmaceutical Industry

The pharmaceutical industry, a critical player in global healthcare, engages in the research, development, manufacturing, and distribution of medicinal products. Operating at the intersection of science and commerce, pharmaceutical companies invest heavily in research and clinical trials to bring innovative drugs to market. Strict regulatory oversight guides their operations, ensuring safety and efficacy. Supply chain management, distribution networks, and partnerships with healthcare providers are integral to efficiently delivering lifesaving medications to patients worldwide. The industry's success hinges on scientific breakthroughs, regulatory compliance, and an intricate balance between public health and business sustainability.

The adoption of digital technologies such as artificial intelligence (AI), machine learning (ML), and advanced analytics is revolutionizing the pharmaceutical industry, accelerating drug discovery, optimizing development, and enhancing patient care. AI-driven platforms like AlphaFold [1] have transformed structural biology by accurately predicting protein folding, significantly reducing the time and cost of screening protein-drug interactions and developing new drug candidates. In parallel, ML algorithms analyze vast datasets to identify promising drug candidates, simulate clinical trials, and optimize formulations. The rise of customizable medicines, including AI-powered precision therapies and mRNA-based treatments, is enabling tailored medical solutions that address individual genetic profiles, improving efficacy and minimizing side effects. Digital health technologies are also enhancing prescription management—AI-driven systems can detect potential drug interactions, predict adverse reactions, and optimize dosing recommendations based on patient history. Additionally, digital supply chain solutions, blockchain for drug traceability, and smart manufacturing processes powered by the Industrial Internet of Things (IIoT) are ensuring higher quality, compliance, and efficiency in pharmaceutical production. As digital transformation continues to proliferate, it is driving unprecedented innovation, making medicine more effective, personalized, and accessible, ultimately reshaping the future of healthcare.

2.6 Agile Business Powered by Digital Transformation

The chemical process and manufacturing industry find themselves at the crossroads of unprecedented change, with three seismic transformations reshaping the landscape: Digitalization, Automation, and Sustainability. In the rest of this chapter, we delve into the intricate tapestry woven by these transformations, exploring their individual nuances, interconnectivity, and the imperative role they play in shaping agile businesses. The journey begins with an overview exploration of each transformation and their unique impact on the industry.

2.6.1 Digital Transformation: Unleashing the Power of Data

Digital transformation in the chemical processing industry involves the integration of advanced technologies such as IoT, data analytics, and cloud computing. These technologies are applied to enhance process monitoring, optimize chemical production, and improve data-driven decision-making. From real-time analysis of reaction kinetics to predictive maintenance of equipment, digital transformation is revolutionizing how chemical processes are monitored and managed. Digital transformation, at its core, is not just about adopting cutting-edge technologies; it is a profound shift in how businesses leverage data. In the chemical process and manufacturing space, where precision and efficiency are paramount, the ability to harness data emerges as a game-changer.

2.6.2 Automation: Bridging the Gap Between IT and OT

Automation plays a pivotal role in the chemical processing industry by implementing technology to control and monitor various aspects of production. From automated batch processing to the use of robotics in hazardous environments, automation increases efficiency and safety. It ensures precise control over chemical reactions, reduces human intervention in risky tasks, and supports the production of high-quality and consistent chemical products. Automation, a linchpin in the pursuit of operational excellence, introduces a fundamental shift in how tasks are executed. Automation enables the capabilities such as streamlining processes, reducing human errors, and optimizing resource utilization. The integration of Operational Technology (OT) and Information Technology (IT) becomes a focal point, emphasizing the need for a cohesive approach to unlock the full potential of automation.

2.6.3 Sustainability: Balancing Environmental and Business Needs

Sustainability initiatives in chemical processing focus on adopting eco-friendly practices, reducing environmental impact, and optimizing resource usage. This includes the incorporation of green chemistry principles, the adoption of renewable energy sources, and the implementation of circular economy concepts. Chemical manufacturers are actively working towards minimizing waste, lowering emissions, and creating products with a reduced environmental footprint. Sustainability is not a buzzword but a business imperative. Simultaneously, the focus shifts to how sustainability strategies fortify businesses, fostering adaptability, rapid innovation, and resilience in the face of disruptions.

2.7 Building a Strong Business Case

Embarking on the transformative journey of digital evolution within the chemical processing industry requires a strategic orchestration. Recognizing the integral role that automation and sustainability play in fortifying the business case is essential. In this narrative, digital innovation stands as our destination, with automation and sustainability acting as steadfast allies not only in environmental responsibility but also in ensuring business continuity and resilience.

2.7.1 Navigating Synergy

Our vision for digital transformation entails not just a technological shift but a synergy with automation and sustainability. Rather than viewing these transformations as parallel endeavors, adopting a collaborative perspective becomes the cornerstone of our business case. This synergy ensures that the efficiencies gained from automation and the principles of holistic sustainability are seamlessly integrated into the fabric of our digital future.

2.7.2 Catalyzing Operational Excellence

Automation takes a pivotal role in catalyzing operational efficiency, becoming a driving force for our journey towards digital transformation. By emphasizing how automation enhances precision, minimizes errors, and enables dynamic scaling, the business case articulates a culture of excellence. This aligns with the overarching vision of achieving operational brilliance through the integration of digital tools while ensuring the sustainability of operations for long-term business continuity.

2.7.3 Quality and Innovation Leadership

Enhancing product quality and fostering innovation are critical aspects of the business case. Advanced analytics and modeling provided by digital tools offer an unprecedented understanding of chemical reactions and processes, enabling the production of high-quality and consistent products. In the journey to elevate product quality and drive innovation, the pivotal role of automation cannot be overstated. Automation not only acts as the linchpin for ensuring precision in chemical manufacturing processes but also lays the foundation for seamless integration with advanced analytics and modeling. By embracing automation, the business not only achieves operational efficiency but also establishes a dynamic framework where real-time data insights can contribute to a deeper understanding of the underlying chemical reactions in a process. By leveraging simulation and virtual prototyping, one can accelerate product development, ensuring a quicker time-to-market and positioning. This interconnectedness amplifies the impact of digital tools, making automation transformation a cornerstone in the quest for unparalleled product quality, swift innovation, and a leadership position in the competitive industry landscape.

2.7.4 Supply Chain Resilience and Responsiveness

Digital transformation offers a significant advantage in supply chain management. From demand forecasting to integrated supply chain visibility, digital technologies provide the tools to optimize inventory, reduce excess stock, and enhance overall supply chain resilience. This not only ensures operational efficiency but also positions the business to respond effectively to market fluctuations and disruptions. Envisioning a digitally transformed supply chain requires acknowledging the collaborative strength of automation and sustainability. Automation's role in efficient inventory management and dynamic scaling aligns seamlessly with the goals of digital transformation. Highlighting this synergy reinforces the business case and emphasizes the vision of a supply chain that is not only resilient but also responsive, ensuring the continuity of operations.

2.7.5 Safety, Compliance, and Reputation Enhancement

Ensuring safety and compliance is a non-negotiable aspect of the business case. Digital tools contribute to risk mitigation by providing real-time monitoring and supporting automated safety protocols. This not only minimizes the likelihood of accidents but also ensures regulatory compliance. The ability to track and document compliance enhances the business's reputation as a responsible and compliant entity, fostering trust among stakeholders. Safety and compliance remain paramount in our vision for a digitally transformed future. Automation, with its real-time monitoring capabilities, becomes indispensable in ensuring safety and compliance. This collaborative effort fortifies the business case, contributing not only to enhanced corporate reputation but also ensuring business sustainability through the responsible and compliant operation of processes.

2.7.6 Holistic Sustainability

In an era where environmental stewardship is a growing concern, the business case for digital transformation in the chemical processing industry cannot overlook sustainability. Resource optimization, waste reduction, and energy efficiency achieved through digital technologies align with eco-friendly practices. Reducing the carbon footprint not only meets societal expectations but also positions the business as a responsible corporate citizen committed to sustainable practices. However, sustainability is positioned not merely as an environmental concern but as a cornerstone for the business to thrive. Automation's role in resource optimization and waste reduction aligns seamlessly with sustainability goals that ensure the long-term viability of the business. By spotlighting this harmonious collaboration, the business case paints a holistic picture of environmental responsibility and business resilience, reinforcing the vision of a sustainable, future-ready organization.

2.7.7 Customer-Centric Innovation

In our pursuit of digital innovation, the customer remains at the core. Here, the synergy between digital tools and automation becomes a driving force for customer-centric innovation. Customization and responsiveness, facilitated by automation, resonate with the vision of being a market leader in meeting customer demands through digital transformation, ensuring not just business sustainability but also customer loyalty. Sustainability, often viewed as an environmental imperative, also emerges as a catalyst for innovation. Integrating sustainability practices into the innovation process can lead to the development of novel solutions and processes. For instance, the imperative to reduce waste may drive the exploration of alternative raw materials or the implementation of closed-loop systems. This holistic approach not only aligns with environmentally conscious practices but also positions the business as an innovator with a forward-thinking and socially responsible ethos which consumer demands.

2.8 Closing Thoughts

Navigating the transformative path of digital evolution in the chemical processing industry requires a comprehensive understanding of the collaborative efforts of automation and sustainability. The business case, therefore, becomes a testament to the interconnectedness of these transformations, reinforcing the vision of a digitally transformed, operationally efficient, sustainable, and resilient future. By acknowledging and leveraging the strengths of automation and sustainability in tandem with digital innovation, we pave the way for a holistic transformation that transcends technological boundaries, ensuring both environmental and business sustainability.

Reference

1. John Jumper, Richard Evans, Alexander Pritzel, Tim Green, Michael Figurnov, Olaf Ronneberger, Kathryn Tunyasuvunakool, Russ Bates, Augustin Žídek, Anna Potapenko, Alex Bridgland, Clemens Meyer, Simon A. A. Kohl, Andrew J. Ballard, Andrew Cowie, Bernardino Romera-Paredes, Stanislaw Nikolov, Rishub Jain, Jonas Adler, Trevor Back, Stig Petersen, David Reiman, Ellen Clancy, Michal Zielinski, Mikolaj Steinegger, Michalina Pacholska, Tomáš Berka, Anna-Friederike Fabian, Simon J. Michael, Helen F. J. Askew, AlphaGo Zero, AlphaZero, AlphaFold, AlphaStar, and AlphaGo. Highly accurate protein structure prediction with alphafold. *Nature*, 596:583–589, 2021.

The Path to Digital Transformation 3

3.1 Evolution of Manufacturing

The story of manufacturing is one of continuous evolution, marked by transformative technological shifts known as industrial revolutions. Each revolution has brought new possibilities, reshaping factories and redefining how goods are produced. The journey from steam-powered machinery to the current era of automation, driven by advanced digital technologies, is a testament to the relentless pursuit of efficiency, precision, and innovation in manufacturing.

The first industrial revolution laid the foundation for modern manufacturing by introducing steam-powered machinery starting from the late 18th century. Steam engines replaced manual labor and water-driven systems, ushering in an era of mechanization. Factories became centers of production, particularly in textiles and mining, as steam- powered machines vastly increased output and efficiency.

The second industrial revolution brought about the widespread adoption of electricity, leading to a significant leap in manufacturing capabilities in the late 19th century. Factories embraced assembly lines and mass production techniques, most famously exemplified by Henry Ford's automotive assembly line. This era witnessed the birth of consumer goods industries and the rise of modern manufacturing plants.

The advent of computers and automation marked the third industrial revolution starting from the mid 20th century. This period saw the integration of electronic controls into manufacturing processes, allowing for greater precision and control. Numerical control (NC) machines and programmable logic controllers (PLCs) automated tasks, enabling higher production volumes and quality consistency. computer-aided design (CAD) and computer-aided manufacturing (CAM) further revolutionized product design and production planning.

The fourth industrial revolution, characterized by the fusion of digital, physical, and biological technologies, is unfolding in the present era. Automation in manufacturing has reached new heights with the integration of smart technologies such as the internet of things

(IoT), artificial intelligence (AI), and advanced robotics. A key driver of this revolution is the interconnectedness of digital systems powered by the internet. Smart factories, equipped with interconnected systems, data analytics, and autonomous machines, exemplify the pinnacle of this revolution.

The fifth industrial revolution is at a nascent stage as this book is nearing publication, where industries are aspiring to become more human centric [1]. This marks a shift from the hyper-automation and AI-driven efficiencies of the fourth industrial revolution to a human-centered approach that integrates advanced technology with empathy, creativity, and sustainability. Unlike previous revolutions that prioritized productivity and digitization, the fifth revolution emphasizes collaboration between humans and intelligent machines to enhance well-being, inclusivity, and ethical responsibility. Key human factors in this transition include job satisfaction, mental and physical well-being, ethical AI development, and the balance between automation and human creativity. This revolution envisions a society where technology serves humanity rather than replaces it, fostering workplaces that prioritize purpose and meaningful engagement. As a result, industries will move toward personalized, ethical, and sustainable production, leading to a world where economic growth aligns with environmental and social welfare, ensuring a more balanced and equitable future.

3.2 Progress of Automation in Manufacturing

Automation in manufacturing has evolved from the mechanization of tasks to the current era of intelligent, data-driven production. Robotics, AI, and IoT have become integral components of modern factories, optimizing efficiency and responsiveness. Automation not only streamlines routine tasks but also enables adaptive manufacturing, where production lines can swiftly adjust to changing demands and customize products with unprecedented precision.

Smart manufacturing is the manifestation of the fourth industrial revolution in factories. Through the seamless integration of digital technologies, smart manufacturing facilitates real-time data exchange, predictive maintenance, and agile production processes. Advanced algorithms analyze vast datasets, enabling machines to make informed decisions, optimize operations, and continuously improve efficiency.

Smart factories leverage cutting-edge information exchange technologies to enhance automation in manufacturing processes. Intelligent machines equipped with sensors, actuators, and connectivity form a networked ecosystem, enabling real-time data exchange and decision-making. This interconnectedness not only optimizes individual processes but also allows for holistic control and synchronization across the entire production chain.

Data analytics plays a pivotal role in the modern smart plant, empowering manufacturers with actionable insights for decision-making. The integration of sensors and IoT devices in manufacturing equipment generates a wealth of real-time data, offering a comprehensive view of production processes. Analyzing this data allows manufacturers to identify bottlenecks, optimize resource allocation, and continuously improve operational efficiency.

The manufacturing sector has witnessed a substantial increase in the adoption of data driven adaptive automation in production processes. Robotics, Artificial Intelligence (AI), and the Internet of Things (IoT) are at the forefront of this transformation. Robots are being employed for repetitive and precision-based tasks, leading to increased accuracy, reduced errors, and enhanced production speed. AI algorithms analyze vast datasets to optimize production schedules, predict maintenance needs, and improve overall efficiency.

One of the key impacts of the fourth industrial revolution on manufacturing is the ability to achieve greater agility, customization, and flexibility in production. Advanced automation technologies enable the seamless adaptation of production lines to varying product specifications and customer demands. This flexibility allows manufacturers to efficiently produce small batches of customized products, responding to market trends and consumer preferences with agility.

The fourth industrial revolution is breaking down silos in manufacturing by integrating the supply chain through digital technologies. Automated systems facilitate seamless communication between suppliers, manufacturers, and distributors, optimizing logistics and reducing lead times. Supply chain integration not only enhances efficiency but also enables a more responsive and agile manufacturing ecosystem capable of adapting to dynamic market conditions.

3.3 Challenges and Opportunities

While the progress of automation in manufacturing presents numerous opportunities, it also poses challenges. Workforce re-skilling, cybersecurity concerns, and the need for substantial investments in technology are considerations that manufacturers must address. However, the potential benefits, including increased productivity, reduced costs, and enhanced product quality, underscore the transformative impact of automation on the manufacturing sector.

From steam-powered machinery to the current era of smart manufacturing, automation has been the driving force behind increased efficiency, precision, and innovation. As we stand on the cusp of the fourth industrial revolution, the future promises a manufacturing landscape where human-machine collaboration and advanced technologies propel the industry into new realms of possibility.

As automation becomes more pervasive, manufacturers must navigate the complexities of technological integration, address workforce challenges, and capitalize on the vast potential for increased efficiency and innovation. Embracing the opportunities presented by the fourth industrial revolution is not just a strategic choice but a necessity for manufacturers aiming to remain competitive in an increasingly digitized and interconnected global marketplace.

3.4 The Case for Digital Transformation

The chemical processing industry stands at the doorstep of a profound transformation driven by digitalization. As technology continues to evolve, the adoption of digital transformation has become more than just a strategic choice—it is now a necessity for the sustainable growth and competitiveness of the chemical industry. In this section, we discuss the driving forces and rationale of digital transformation and explore how it can bring transformative benefits to the chemical processing sector.

While dealing with the subject of digital transformation in the manufacturing environment, we have particularly focussed on the chemical processing industry in this book. This is perhaps attributable to the authors' specific backgrounds and professional experiences. However, this perspective largely stems from the special needs and considerations pertinent to chemical processing plants that pose a unique set of challenges with respect to implementing digital transformation in these plants. Among various industrial manufacturing environments, we feel that many segments of the chemical processing industry are either underserved or have nuances that have not been considered in general digitalization initiatives.

The processes in a chemical plant are not what we traditionally attribute to conventional discrete manufacturing principles subject to robotic processes common in many other segments of the manufacturing industry. While manufacturing a molded solid object, a semiconductor chip, an LED screen, or a car involves tasks that can be automated in a manufacturing environment through application of vision and controlled mechanical motion, the scenario is considerably diverse in chemical processing. Even the simple task of monitoring involves a plethora of sensing mechanisms, as it is non-trivial to visually monitor the motion of liquids inside a pipe or a pressure vessel. Chemical processes involve dealing with multiple states of matter (solids, liquids, and gases) that can spontaneously form within any of the process operations, and we may need to measure phase changes, and composition (concentrations) in each phase. The translation of the physical measurements of composition or phase boundaries to a cyber-realm involves more complex transducers and measurement devices.

Chemical processing is also quite non-linear in time–whether the processes involve blending, mixing, separation, or reactions. Therefore, connecting the physical factory environment to the cyberspace, including implementation of sensors, monitoring a process, or analysis of the results can be quite diverse and fragmented for the CPI. A single unique recipe generally does not apply to implementing a digital transformation paradigm in different segments of the industry.

Accordingly, the technology implementation costs, required skillsets, timelines, and the return on investment on such implementations can be quite varied. There are only a handful of existing technical resources that specifically focus on topics pertinent to implementation of digital transformation for the CPI. This book aims to cover such topics, sometimes straying

3.4 The Case for Digital Transformation

from the conventional emphasis on principles of robotic manufacturing in other segments of the manufacturing industry.

With this in mind, we outline below a few pertinent considerations that are generally thought of as key drivers for digital transformation initiatives in the manufacturing sector.

3.4.1 Operational Efficiency and Optimization

Digital transformation empowers chemical processing plants to optimize their operations with unprecedented precision. Through the integration of advanced process control systems, real-time data analytics, learning and intelligent algorithms, manufacturers can streamline production processes, minimize waste, and enhance overall operational efficiency. Automated monitoring and control systems enable continuous process improvements, reducing downtime and increasing yield.

3.4.2 Predictive Maintenance and Asset Management

The adoption of digital technologies allows chemical processing plants to transition from reactive to proactive maintenance strategies. Predictive maintenance, enabled by sensors and data analytics, helps anticipate equipment failures before they occur. This not only extends the lifespan of critical assets but also minimizes unplanned downtime, ensuring that production processes remain uninterrupted.

3.4.3 Supply Chain Visibility and Collaboration

Digital transformation provides end-to-end visibility into the supply chain, from raw material sourcing to product delivery. Through the use of advanced analytics and interconnected systems, chemical manufacturers can optimize inventory management, track shipments in real-time, and collaborate seamlessly with suppliers and distributors. This transparency enhances supply chain resilience and responsiveness to market fluctuations.

3.4.4 Quality Control and Compliance

Digital technologies offer robust solutions for quality control and regulatory compliance in the chemical processing industry. Automated systems monitor and analyze production data, ensuring that products meet stringent quality standards. Additionally, digital documentation and reporting tools streamline compliance processes, reducing the risk of errors and facilitating adherence to environmental, health, and safety regulations.

3.4.5 Innovation and Product Development

Digital transformation serves as a catalyst for innovation in the chemical industry. Virtual simulations, modeling, and data-driven insights enable faster and more cost-effective product development. Advanced analytics help identify market trends, allowing manufacturers to respond swiftly to changing consumer demands and stay ahead of the competition.

3.4.6 Energy Efficiency and Sustainability

The chemical processing industry faces increasing pressure to reduce its environmental impact. Digital transformation plays a pivotal role in achieving sustainability goals by optimizing energy consumption, reducing emissions, and promoting resource efficiency. Smart manufacturing processes, enabled by digital technologies, contribute to a more eco-friendly and sustainable industry.

3.4.7 Risk Management

Digital transformation plays a crucial role in knowledge retention by systematically capturing, storing, and disseminating expertise within an organization, reducing the risks associated with employee attrition. Technologies such as cloud-based knowledge management systems, AI-powered documentation tools, and digital twins ensure that critical process knowledge, best practices, and troubleshooting insights are preserved in a structured, accessible format. Interactive training platforms and augmented reality (AR) applications further enhance knowledge transfer by providing immersive, hands-on learning experiences for new employees. Additionally, AI-driven analytics can identify gaps in workforce expertise and recommend targeted training programs to upskill personnel. By digitizing institutional knowledge and making it easily accessible, organizations can maintain operational continuity, reduce dependency on individual expertise, and ensure a seamless transition when experienced employees retire or leave the company.

The interconnected nature of digital systems necessitates robust risk management and cybersecurity measures. Digital transformation involves implementing state-of-the-art cybersecurity protocols to protect sensitive data and critical infrastructure. Proactive risk management strategies, including scenario analysis and contingency planning, ensure the resilience of chemical processing plants in the face of potential threats.

3.5 Is Digital Transformation a Revolution in the CPI?

The CPI has long been at the forefront of automation and digitalization, particularly in advanced sectors like petrochemical refining, energy generation, and pharmaceutical manufacturing. Even before the emergence of Industrie 4.0, these industries had deployed sophisticated distributed control systems (DCS), supervisory control and data acquisition (SCADA) systems, and programmable logic controllers (PLCs) to manage complex chemical processes with minimal human intervention. Furthermore, in-line process analyzers and advanced sensor networks enabled real-time data acquisition, contributing to extensive process history records and institutional knowledge retention. Given this high degree of automation and digital control, the question arises: What additional value does Industrie 4.0 bring to an already digitized industry?

While traditional DCS and SCADA systems provided real-time monitoring and control, they lacked the ability to leverage big data analytics, machine learning (ML), and artificial intelligence (AI) for predictive decision-making. Industrie 4.0 introduces AI-driven process optimization, anomaly detection, and predictive maintenance strategies that go beyond simple trend analysis and alarm systems. By integrating digital twins, chemical plants can now simulate process variations in real-time, optimizing production parameters dynamically rather than relying on static set-points determined through historical trial and error.

Traditional automation in CPI relied on predefined control loops and operator expertise for optimization, but Industrie 4.0 enables self-learning and self-optimizing process control. Reinforcement learning algorithms can adjust operational conditions in response to fluctuations in raw material quality, energy prices, and environmental conditions, leading to more resilient and cost-effective plant operations. Closed-loop AI-based control can adjust reaction conditions, feedstock ratios, or catalyst loading based on real-time yield predictions, significantly improving process efficiency.

A key limitation of legacy digital control systems is their siloed architecture, where process control data is often confined to plant-level SCADA systems, requiring manual interventions for supply chain adjustments. Industrie 4.0 brings Industrial IoT (IIoT) and cloud-based connectivity, allowing seamless integration between process control, enterprise resource planning (ERP), supply chain management, and even customer demand forecasting. This vertical and horizontal integration enables more adaptive production scheduling, reducing waste and improving responsiveness to market changes.

While pre-Industrie 4.0 digital automation focused primarily on process reliability and safety, modern plants face increasing threats from cyber-attacks, data breaches, and digital sabotage. Industrie 4.0 introduces advanced authentication, AI-driven threat detection, and secure cloud architectures to safeguard critical process data and prevent unauthorized access to industrial networks. Given the increasing frequency of cyber threats targeting energy and petrochemical infrastructure, these enhancements are essential for ensuring operational continuity.

Despite the extensive automation in pre-Industrie 4.0 plants, operator experience and institutional knowledge remained critical for troubleshooting complex scenarios. Industrie 4.0 leverages augmented reality (AR), digital twin simulations, and AI-driven knowledge management systems to preserve and transfer expert knowledge. Remote monitoring and AI-guided operator assistance systems can bridge the expertise gap, particularly as the industry faces an aging workforce and increasing labor shortages.

While it is true that the CPI was highly digitized before the advent of Industrie 4.0, the new wave of technologies offers a qualitative leap beyond traditional automation. This enables a shift from reactive to predictive operations, from siloed control to integrated ecosystems, and from human-dependent optimization to AI-driven intelligence. These advancements are critical not just for improving efficiency but also for maintaining global competitiveness, sustainability, and resilience in a rapidly evolving industrial landscape.

3.6 Rationale and Timeline of Implementing Digitalization

3.6.1 Process Plant Life Cycle and Digitalization

Figure 3.1 depicts the typical stages of a plant life cycle. In this, the pre-commissioning stages involve planning, design, construction, building and integration. After the plant is commissioned, it gets into the operating phase of its life cycle, spanning one to three decades, during which the plant cycles between production and maintenance modes. As the plant ages, it required replacement of components, de-bottlenecking, and retrofits. Eventually, at the end of its useful life, the plant is decommissioned.

Digitalization plays a transformative role in the entire life cycle of a process plant, influencing everything from initial design and commissioning to long-term operations and optimization. The approach to integrating digital technologies can vary significantly depending on whether digitalization is embedded during the pre-commissioning phase—when the plant is still in the design and construction stages—or implemented as a retrofit in the post-commissioning operating phase. Each strategy offers distinct advantages and challenges, requiring a careful evaluation of feasibility, cost-effectiveness, and expected performance improvements. Table 3.1 summarizes the features, scope, benefits, and risks of greenfield (pre-commissioning) and brownfield (post-commissioning) digital transformation of a process plant.

3.6.2 Pre-commissioning Digitalization

Integrating digitalization in the pre-commissioning phase provides the most seamless and efficient implementation, allowing digital technologies to be embedded into the plant's infrastructure from the outset. This "digital by design" approach ensures that modern tech-

3.6 Rationale and Timeline of Implementing Digitalization

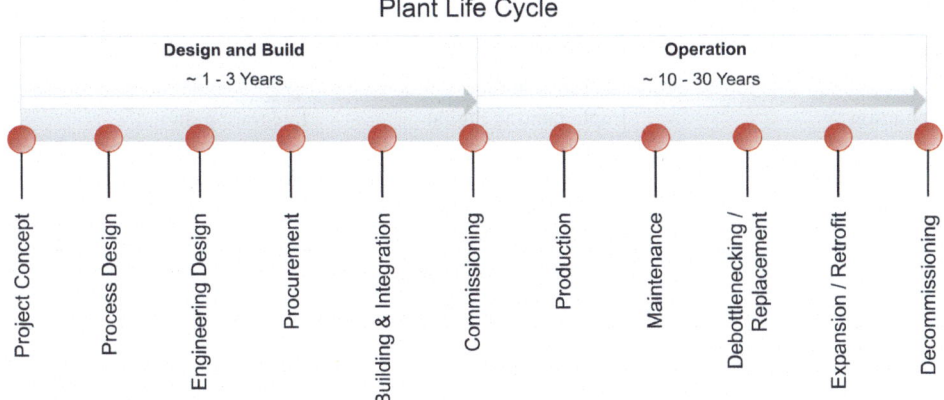

Fig. 3.1 Life cycle of a chemical process plant from project conception to decommissioning. The timeline has a pre-commissioning phase where the plant is conceived, designed and built. Following commissioning, the plant is in operation mode. A digital transformation project can be conceived at any point in the life cycle of a plant. If included in the blueprint stage, the process can be considered as a greenfield digitalization. If digitalization is envisioned after plant commissioning, it is considered retrofit digitalization

nologies such as digital twins, advanced process control (APC), predictive maintenance algorithms, and AI-driven decision support systems are structurally integrated into plant operations rather than being added later. The key rationale and justification of embedding digitalization in today's greenefield plant design can be summarized as:

Seamless Integration with Plant Design Digital models can be developed alongside the plant's engineering blueprints, ensuring optimal sensor placement, data connectivity, and process automation from the start.

Digital Twin Development for Virtual Commissioning A digital twin can be built in parallel with the plant design, enabling virtual commissioning, process validation, and operator training before actual startup.

Optimized Control Architecture The entire DCS, SCADA, and Industrial IoT (IIoT) systems can be designed for full interoperability rather than requiring integration with legacy systems.

Early Detection of Design Flaws Using simulation tools and AI-driven predictive analysis, potential bottlenecks, inefficiencies, or safety risks can be identified and addressed before physical construction.

Cybersecurity Integration Security frameworks can be natively built into the digital infrastructure, reducing vulnerabilities associated with retrofitting cybersecurity into older systems.

Table 3.1 Comparison of greenfield and brownfield digital transformation approaches

Aspect	Greenfield digital transformation	Brownfield digital transformation
Definition	Implementing digital transformation in a newly built plant designed from the ground up for smart operations	Integrating digital transformation into an existing plant with legacy systems and infrastructure
Infrastructure	Modern, optimized OT-IT architecture with cutting-edge technologies, designed for high interoperability	Requires retrofitting digital technologies into legacy control systems and infrastructure
Flexibility	High flexibility to adopt best-in-class standards, architectures, and cybersecurity frameworks	Limited by existing equipment, protocols, and operational constraints
Cost	High initial capital expenditure, but optimized long-term costs due to efficient design	Lower upfront costs but potential hidden costs due to compatibility challenges and necessary upgrades
Implementation Complexity	Easier to design seamless integration of digital tools and infrastructure	Complex, requiring extensive planning for phased upgrades, data migration, and system compatibility
Time to Deploy	Longer planning and construction time, but fewer integration challenges post-deployment	Faster deployment as it builds on existing infrastructure, though risk of operational disruptions exists
Value Proposition	Enables the most advanced form of automation, intelligence, and sustainability from inception	Allows legacy systems to evolve into intelligent, connected systems without disrupting ongoing production
Risks	High initial risk due to large investments and potential delays in construction	Risks of downtime, incompatibility, and resistance to change from existing workforce
Scalability	Designed for scalability with cloud, edge computing, and IIoT integration	Scalability depends on legacy system constraints and upgrade potential

Cost Savings in Long-Term Operations Since the plant is designed to operate with digital optimization from day one, long-term savings arise from reduced downtime, improved process efficiency, and lower maintenance costs.

3.6 Rationale and Timeline of Implementing Digitalization

Pre-Commissioning Digitalization allows

- Selection of appropriate sensor technologies and communication protocols for seamless data collection.
- Integration of digital twin and virtual plant simulation tools to ensure real-world accuracy in predictive modeling.
- Cybersecurity design to mitigate risks from the outset.
- Scalability of digital solutions to accommodate future expansions or modifications.
- Training programs for operators and engineers using immersive AR/VR-based simulation environments.

3.6.3 Retrofit or Post-commissioning Digitalization

For existing process plants that were designed before the Industrie 4.0 era, retrofit digitalization allows plants to incorporate modern technologies into their legacy infrastructure. While this approach can be more complex due to compatibility challenges, it enables incremental upgrades without requiring a complete overhaul of the facility. Although more niche and requiring considerable effort and customization for ageing plants, and even costlier to implement in some cases, retrofit digitalization can be rationalized for some older plants due to the following reasons:

Extending Equipment Life and Process Optimization By adding IIoT sensors, AI-driven analytics, and cloud-based monitoring systems, aging process equipment can be enhanced to improve efficiency, reduce energy consumption, and extend operational life.

Cost-Effective Incremental Upgrades Instead of a large-scale capital investment, retrofit digitalization allows companies to gradually adopt digital tools in phases, targeting the most critical areas first.

Predictive Maintenance and Operational Efficiency Advanced data analytics can be integrated into legacy systems to provide predictive maintenance insights, reducing unplanned shutdowns and improving asset reliability.

Integration with Modern Supply Chains Retrofit digitalization enables connectivity with modern ERP and supply chain management systems, enhancing responsiveness to fluctuations in demand and raw material availability.

Compliance with Environmental and Safety Regulations Retrofitting digital technologies such as emissions monitoring, leak detection, and automated safety interlocks helps plants meet evolving regulatory standards without significant structural changes.

Cybersecurity Hardening for Older Systems Many older process plants were not designed with modern cybersecurity threats in mind. Retrofitting security layers, such as

AI-driven intrusion detection and blockchain-based data integrity, helps mitigate these risks.

Retrofit digitalization needs to be consiered on a case by case basis, with attention to the state of the older plant, remaining lifetime of the plant, competitiveness of the processes utilized and implemented in the plant, obsolescence of the equipment and control architecture, and overall profitability of the plant. Notwithstanding, exploring retrofit digitalization of existing plants, when implemented judiciously may be justified owing to the following key considerations:

- Compatibility with existing control systems (DCS, PLCs, SCADA) and protocols.
- Scalability and modularity of digital upgrades, ensuring staged implementation without disrupting plant operations.
- Edge computing and cloud integration to enable real-time analytics without requiring full system overhauls.
- Interoperability of added sensors and analytics tools with legacy automation architectures.
- Cybersecurity measures to protect against vulnerabilities in older systems.
- Cost-benefit analysis of digital upgrades versus potential ROI from efficiency gains and reduced downtime.

Both pre-commissioning digitalization and retrofit digitalization provide significant advantages, but their implementation strategies differ based on feasibility, cost, and operational goals. Greenfield projects benefit from seamless digital integration, optimized control architectures, and predictive analytics from the start, while existing plants can incrementally adopt digital tools to enhance operational efficiency, asset reliability, and regulatory compliance. The choice between these approaches depends on factors such as capital investment capacity, existing infrastructure compatibility, and long-term strategic objectives. In practice, most industries adopt a hybrid strategy, where new plants are designed with full digital integration while older facilities undergo targeted digital retrofits to stay competitive in an increasingly data-driven industrial landscape.

3.7 Outlook

In conclusion, the digital transformation of the chemical processing industry is not merely an option—it is an essential journey towards ensuring long-term competitiveness, sustainability, and innovation. From operational optimization to sustainable practices and innovation, the benefits of embracing digital transformation are transformative, positioning the chemical industry at the forefront of the digital era. As chemical manufacturers embark on this digital journey, they are not only future-proofing their operations but also contributing to the broader evolution of the industry towards a more connected, efficient, and sustainable future.

Reference

1. Praveen Kumar Reddy Maddikunta, Quoc-Viet Pham, Prabadevi B, N Deepa, Kapal Dev, Thippa Reddy Gadekallu, Rukhsana Ruby, and Madhusanka Liyanage. Industry 5.0: A survey on enabling technologies and potential applications. *Journal of Industrial Information Integration*, 26:100257, 2022.

Digital Transformation–Technical Foundations 4

4.1 Introduction

In the previous chapters we discussed the pertinent considerations that provide a rationale for the implementation of digitalization in the process industry. This was done through a consideration of the macro-scale business cases (Chap. 2) as well as some of the macro-scale techno-economic considerations (Chap. 3). In this chapter, we look more closely at the technical considerations required to implement modern digitalization concepts in a project within the chemical process industry. To achieve this, we will first focus on the key vocabulary used in digital transformation and Industrie 4.0, as well as our interpretation of these in context of the CPI. In the previous three chapters, considerable liberty was taken in using a large vocabulary from the fields of digitalization, IT, computational algorithms, and process automation, with a tacit assumption that the reader will be familiar with those terms and their implications. It is possible that a section of the readers will be familiar with those terminologies and their implications, and in that case, a lot of the content of this chapter will most probably seem elementary and repetitive. Such readers may very well skip this chapter, without losing continuity. However, this chapter is aimed for a reader who is not familiar with digital transformation and is seeking a fast-track primer on the topic that technical managers in the process industry can use and modify, as needed, to implement a digitalization strategy into their managed process operations, organizations, and industry. Therefore, a considerable emphasis in this chapter is placed on defining key terms and concepts critical to the implementation of digitalization.

4.2 Technical Premise of Digital Transformation

The key benefits of converting a conventional process plant into a smart plant are circular economic: achieve greater process reliability, optimize operational flexibility, and develop pathways for organization-wide sustainability. The technical basis of achieving this transformation are through digital transformation and automation of the workflows in the process plant. When implementing a digitalization project, one has to minimally demonstrate that the transformation achieves lower production and operating costs, increased process throughput, enhanced remote operability and monitoring leading to less human intrusion into the hazardous production environments, autonomous control, and improved reliability.

A successful digital transformation takes information from a conventional process plant, converts it to digitized information (data), processes the data to develop knowledge, i.e., processed information structured in some way to disseminate contextual and causal information, and then synthesizes the knowledge to deliver wisdom, i.e., useful knowledge that imparts insight and foresight, eventually converting the conventional plant to a smart plant. In this chapter, we map the broad steps of a transformational journey from conventional plant process control to a smart plant automation paradigm.

At the outset, we define two key terms, namely, *digitization* and *digitalization*. These terms are closely related but have distinct meanings:

Digitization refers to the process of converting analog information into a digital format. It is a technical transformation where physical or analog data (such as handwritten documents, printed reports, or analog signals) is encoded into a machine-readable digital format. This is a fundamental step toward computerized and digital information processing. Examples of digitization include scanning a paper document to create a PDF, use of optical character recognition technologies to convert handwritten laboratory records into a digital database, and using sensors to transform physical or analog signals, such as temperature and pressure, into digital data.

The key focus in digitization is to change the data type from physical/analog to digital (binary, hexadecimal, or any computer memory compatible format). However, no fundamental change in how the data is used is envisioned in digitization.

Digitalization refers to the use of digital technologies to improve business processes, enhance decision-making, and create new operational efficiencies. It involves integrating digital tools into existing workflows to optimize or transform how a system, business, or industry operates. Examples of digitalization can include deploying real-time process monitoring dashboards using Industrial IoT (IIoT) that aggregate information (data) from sensors, implementing an AI-driven predictive maintenance system in a chemical plant, or replacing manual paper-based quality control checks with automated digital inspections.

The key focus during digitalization is on process improvement and efficiency gains utilizing digital tools, transforming workflows to leverage digital capabilities. In summary, digitalization is not just about converting data from analog to digital, but also using it intelligently for optimization and automation. In this respect, digitization is an essential component (or the first enabling step) of digitalization.

As a technical manager in a mainstream process industry sector, one's current thinking may be that digitalization of the process plant is solely the responsibility and premise of the chief information officer (CIO) and the Information technology (IT) department. This is certainly not true! In fact, the expertise of the IT professionals will be to create the digitization pathways and formalize the creation of the digital databases as well as the information and communication channels to share and transport such data. The actual digitalization initiatives are built on the foundation of that digitized information and communication platform. The successful digital transformation of any process plant relies heavily on the participation of the process experts during implementation, testing, validation, and utilization cycle of the project. Furthermore, it may be argued that the process expert's role is a pivotal one in digital transformation, because such an expert defines and measures its success in transforming the operation and the industry. We have often seen that digitalization strategies that include, or often place such technical domain experts into leadership roles in the teams implementing digital transformation lead to measurable success. Such experts can lay out the necessities that are domain specific, provide insight to the teams about what information is sought, how the knowledge-base can be put to use, and essentially provide the experiential wisdom that is eventually embedded in the software tasked to provide the revolutionary leap in making the factory a smart factory.

4.3 The Industrie 4.0 Framework

In this chapter we chose to use the framework and vision of *Industrie 4.0*, first introduced in 2011 at the Hannover Messe (Hannover Trade Fair) in Germany, as part of a strategic initiative by the German government to enhance manufacturing competitiveness. Industrie 4.0 (also written as Industry 4.0) is the new wave of industrial digitalization which has garnered considerable attention since 2013 when it was formalized in a report [1] as the "fourth industrial revolution", characterized by the integration of cyber-physical systems (CPS), the Internet of Things (IoT), artificial intelligence (AI), cloud computing, and advanced automation into industrial processes. The concept represents a paradigm shift in manufacturing and industrial production, where digitalization and connectivity enable smart factories that operate with increased efficiency, flexibility, and intelligence. It is quite remarkable that while Industrie 4.0 is neither a manifesto, directive, or a regulatory framework, it has been adopted worldwide as a guiding framework and vision for implementing digital transformation.

Industrie 4.0 provides a facile and general framework of understanding digitalization and smart plants, primarily in industrial manufacturing sectors. However, the vision and framework transcend industrial processes and have even been adopted in urban and societal planning. At the heart of Industrie 4.0 is the concept of a smart factory. A smart factory adopts information and communication technology into its instrumentation and control networks to evolve manufacturing to a higher level of automation, adaptability, and responsiveness. Smart

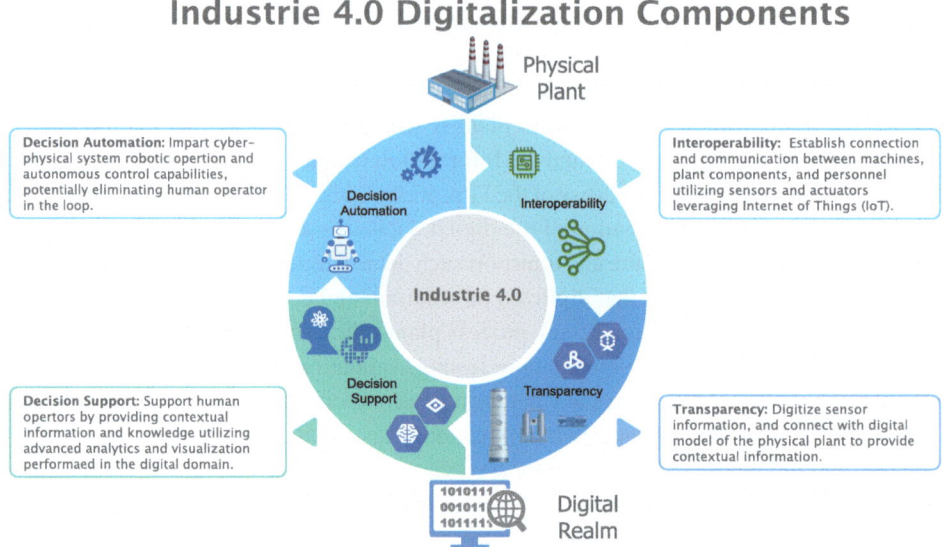

Fig. 4.1 Schematic digital transformation framework representing Industrie 4.0 design philosophy and the closed physical-digital-physical (PDP) loop

machines are automatons that self-optimize, self-configure, and self-learn. These machines can sense, adjust, adapt, and learn, utilizing machine learning (ML) combined with artificial intelligence (AI) to even perform tasks without human intervention. The anticipated end result is superior cost efficiencies, reliable operations, and improved quality of services, as well as reducing the involvement of human operators from tending machines in hazardous operating environments.

Industrie 4.0 is based on four fundamental design principles, namely, interoperability, Information transparency, decision support, and decision automation. Figure 4.1 schematically depicts the interaction between these four design principles to create a closed loop cyber-physical system.

Interoperability is the ability of machines, devices, sensors, and people to connect and communicate with each other, preferably through the Internet of Things (IoT).

Information transparency is the ability of information systems to create a virtual copy of the physical plant, enriching the digital plant models with sensor data. It requires the aggregation of raw sensor data to higher-value, contextual information.

Decision support is technical assistance to human operators or service personnel. First, it provides automation systems the ability to support operators and engineers by aggregating and visualizing information contextually and comprehensively for making informed decisions and solving urgent problems on short notice. Second, it physically supports

4.3 The Industrie 4.0 Framework

operators by conducting a range of tasks that are unpleasant, repetitive, too exhausting, or unsafe for humans.

Decision automation is the capability of plants to make operational decisions and to perform their tasks autonomously. Only during serious exceptions, interferences, or conflicting goals, are tasks delegated to a higher level for human intervention.

Industrie 4.0 leverages the Internet of Things (IoT)—a network of assets and devices (things) embedded with sensor and communications technology, that allows data exchange with other devices over the internet. These are cyber-physical systems such as sensors, actuators, vision systems, and robotics to collect plant information that can be used in near real-time by manufacturers to increase production efficiency.

The primary drivers for a process industry to implement fast track digitalization are the principles of Industrie 4.0, together with an understanding of two "loops:" the *Physical to Digital to Physical (PDP)* loop, and the *Data, Information, Knowledge, and Wisdom (DIKW)* loop. The core technical implementation of digitalization centers around the PDP and DIKW loops.

4.3.1 The PDP Loop

The PDP loop—also shown in Fig. 4.1—is the pathway connecting the physical plant to its digitalized counterpart (the digital domain). Data flows both from the plant to the digital realm as well as from the digital realm to the physical plant when the loop is closed. First the loop collects and transports data from the physical plant to the digital domain (P to D). The data is usually collected from the plant by the sensors that collect the physical information and convert this to digitized data. The digitized is contextualized and stored in databases, that is connected with the automation software in the digital realm. These digital models perform analyses at various levels of complexity, e.g., basic statistical analysis, correlations, predictive analytics, machine learning, or even comparing the plant data with its physics-based mathematical model (the digital twin). After analysis, the assimilated information in the digital realm is transferred back to the physical plant (D to P) for actions to optimize and enhance the performance of the plant. In human operated plants, the digital to physical information transfer is often done by humans in the loop. Such a system is referred to as human in the loop (HITL) system. In more advanced plants, the human interactions could be excluded from the loop, allowing completely autonomous operation of the plant through advanced robotics or process control. Advancements in communications protocols and technology, computation, big data, augmented reality, and powerful analytics make the PDP loop a reality, and its implementation economics well within the reach of many industries.

4.3.2 The DIKW Loop

The DIKW pyramid (Data-Information-Knowledge-Wisdom) is a hierarchical model describing how raw data transforms into actionable wisdom. The origins of this concept can be traced back to Russell Ackoff [2] in the context of systems thinking and organizational learning, though earlier discussions appeared in knowledge management and information science literature. The model was later refined and widely applied in decision-making, artificial intelligence, and digital transformation. In digital transformation system design, the DIKW framework is crucial for structuring data flows—from collecting raw sensor data (D) in industrial settings, to processing it into meaningful information (I), deriving knowledge (K) through analytics, and ultimately supporting wisdom (W) in autonomous decision-making and optimization strategies. This model underpins the Industrial Internet of Things (IIoT), AI-driven process optimization, and predictive maintenance, ensuring that digital transformation initiatives lead to actionable insights rather than mere data accumulation.

In early literature, data was represented as the lowest form entity in the pyramidal hierarchy, perhaps representing values and numbers without context. Information is considered as data with context. There are also cases, when the delineation and hierarchy between data and information is fuzzy. Whether data comes first or information comes first has remained fuzzy, with data sometimes being represented as a digitized version of information. For instance, a number 25.0 without any unit is a digital word or a few bytes of data. However, put in context with a unit, such as Celsius, it represents temperature, while in context of a unit of kilometers/hour, it represents speed. The unit puts into perspective an information that is meaningful to the purveyor on the pure number of the data. In context of digitization and information technology applications, the interpretation of information and data can be altered. Information is often interpreted as the physical quantity or variable measured using a sensor, and data can be the digital realm representation of that information. Therefore, a more rational interpretation of data and information may seem to be that information is physical, while data is the digital domain representation. In this regard, it is not a sacrilege to redefine the DIKW cycle as IDKW (information, data, knowledge and wisdom) in context of digitalization.

Notwithstanding the fuzziness in considering the relative hierarchy of information and data, it is important to note that the words knowledge and wisdom always refer to somewhat higher levels of abstraction of information. Knowledge is often an outcome of a scientific rationalization process, whereby we define a problem, create a hypothesis (a guess to the probable solution to the problem), collect data or information to study the problem, analyze the data to prove or disprove the hypothesis, and arrive at a conclusion. Information processed through this systematic approach, particularly the conclusion drawn, forms a nugget of knowledge, which can itself be another piece of information. Finally, wisdom is often attributable to experiential knowledge of an expert, gleaned from long-term pursuit of a specific type of problem, observing a plethora of data and information, and gleaning multiple nuggets of knowledge from these endeavors. Wisdom, at least in the human context,

4.3 The Industrie 4.0 Framework

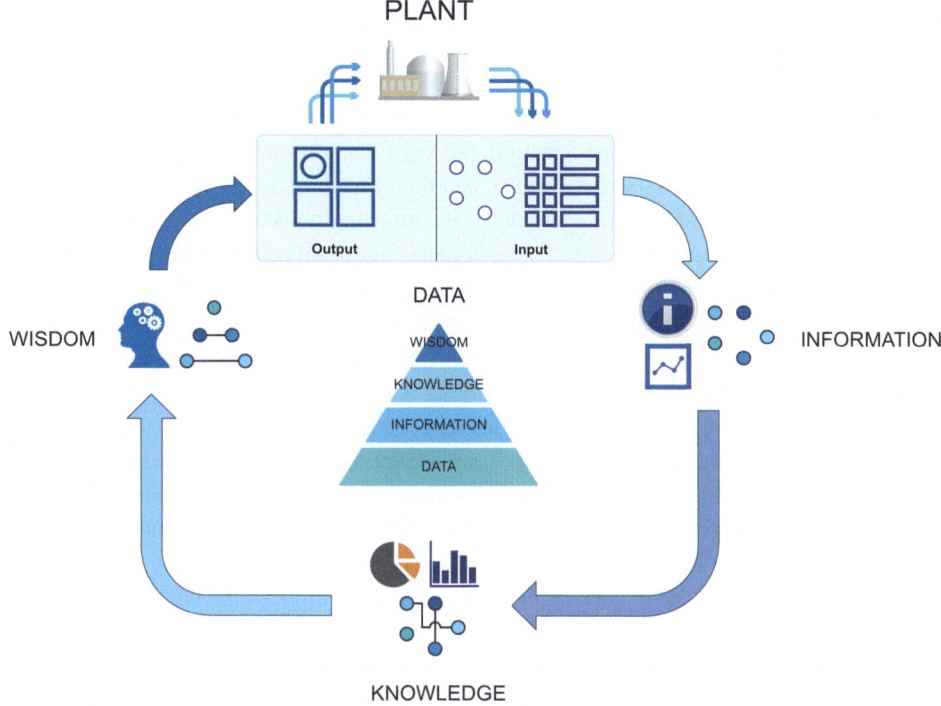

Fig. 4.2 The data, information, knowledge, and wisdom (DIKW) loop

is often associated with deep learning and thinking, as well as experience over a long time, attributable to so-called experts.

There have been re-imaginations of the DIKW pyramid in literature, with some representations of the relationship between these four terms shown as a chain or a loop. In this chapter, we consider the DIKW conversion process as a cycle, without attributing any hierarchy to the four terms: data, information, knowledge, and wisdom. DIKW, used in the context of digital transformation, is the data, information, knowledge, and wisdom loop. A successful digital transformation entails closure of a DIKW loop. A simple figure describing the loop is shown in Fig. 4.2.

The physical state of a process plant or factory is provided by sensors monitoring the plant, and are converted to *data* in a database. This database is often referred to as the plant historian. Historians predominantly store data as time series data, implying that the data are recorded at fixed time intervals. Some state information are also stored as asynchronous data where the information are not collected at regular intervals. Finally, there may be other properties of the plant that can be stored in the form of "meta-data", information that describes the fixed physical properties and parameters of the equipment (such as dimensions, rated power of motors, etc). The combination of these types of data becomes pertinent *information* for stakeholders with relevant expertise and experience. One set of data can be information relevant to the plant operators, whereas another set of data becomes relevant for the process specialist or the product quality assurance team, yet another set of information may be relevant to the production manager who is monitoring the plant uptime, capacity utilization, and energy intensity. With digital transformation, the key focus is to aggregate all data from the plant into a single database (source of truth), and then disseminate the relevant information to the appropriate stakeholder. Conventionally, the information is translated into *knowledge* when it is processed by specific expert stakeholders. In digital transformation projects, the goal is often to embed some type of algorithmic process that will autonomously process the information into knowledge. In a digitally transformed plant, this is achieved through appropriate digital realm computational algorithms. Finally, the knowledge is transformed into *wisdom*, which can be considered as a higher level intellectual function that provides insight and foresight regarding the plant. This has been the age-old relationship between a factory and it's human operators, managers, and stakeholders. In the context of digitalization, the human tasks of monitoring plant information and converting the information into knowledge and wisdom to operate the plant efficiently and reliably is attributed to a digital computational system (the cyber-system).

It is worth noting that the means of communicating the output of the wisdom back to the plant for a cyber physical system is through another set of data (the output layer data). This output data can be quite different from the input data, and generally represents some higher level abstraction of the historical data.

The DIKW loop delineates the steps needed to convert data from a physical plant into information, knowledge, and wisdom to enable predictive and intelligent automation. It represents the core philosophy of how we define the actions manifested in every PDP loop. For instance, information about temperature in a plant is captured by a thermocouple, the electronics behind which transforms the electrical signals from the thermocouple to a digitized signal that represents data. The data can be transferred over the internet to a cloud-located remote database and stored in a computer memory. Conversion of the information stored as data into knowledge occurs when someone queries the data and processes the information to provide correlations, predictions, classification, or various other "cause and effect relationships" as bits and pieces of knowledge that can be extracted from the data. This can include performing mathematical exercises on historical time series data to draw trend lines. In an autonomously operated digital world, these calculations are often relegated to computers performing these calculations autonomously at fixed intervals, or triggered by other inputs, without the requirement of human intervention. The final step of information processing is to assimilate the knowledge into wisdom. Wisdom is the ability to synthesize knowledge to achieve higher level intelligence such as foresight, predictive cognition, and ability to optimize a process. This wisdom, or ability to tell what could happen in the future, can then guide subsequent operation of the physical plant.

> ***The Evolution of Autonomous Vehicle Technology***
>
> *A compelling example of digitalization's transformative power lies in the evolution of vehicle control systems from basic cruise control to fully autonomous driving with collision avoidance capabilities. In a traditional cruise control system, the vehicle maintains a fixed speed set by the driver, but it lacks awareness of its surroundings. It does not account for other vehicles, obstacles, or sudden changes in traffic conditions. The responsibility of adjusting speed, braking, or maneuvering remains entirely with the human driver. This represents an older paradigm of automation—one that is reactive and dependent on human intervention for complex decision-making.*
>
> *In contrast, a modern autonomous vehicle operates within a completely different paradigm, where it continuously processes real-time data from cameras, lidar, and radar sensors to map its surroundings. The system not only detects objects but also tracks their distance, speed, and projected trajectories at high frequencies. This transformation—from raw sensory input (data) to structured information about nearby vehicles, pedestrians, and road conditions—is the foundation of situational awareness in smart systems. However, merely knowing where objects are is insufficient; the system must also assess whether those objects pose a risk of collision. This transition—from information to knowledge—requires advanced computational models that analyze movement patterns and predict potential hazards.*
>
> *To elevate this knowledge into actionable wisdom, the autonomous vehicle must determine the appropriate response to avoid an accident. If a potential collision is detected, the system can issue an alert to the driver (decision support) or take direct control of the vehicle by braking or steering away from the obstacle (decision automation). For example, a backup alert that produces warning beeps when a driver is reversing near an obstacle is a form of decision support, aiding the human in making better driving choices. However, in fully autonomous collision avoidance, the vehicle applies the brakes on its own, even if the driver is distracted. Similarly, self-parking technology represents an advanced application of decision automation, where the car identifies an available parking space and maneuvers itself into position without human assistance.*

4.4 Components of Digital Transformation

4.4.1 Industrial Control System (ICS)

A plant automation framework is often referred to as the industrial control system (ICS) [3]. It is also referred to as the distributed control system (DCS), or operations technology (OT).

Most plant automation hardware consist of sensors (e.g., gauges, indicators, and transducers for measuring temperature, pressure, levels, or flow) and actuators (pumps, valves, motors, solenoids, switches, etc.)–devices whose operations can be automated by a con-

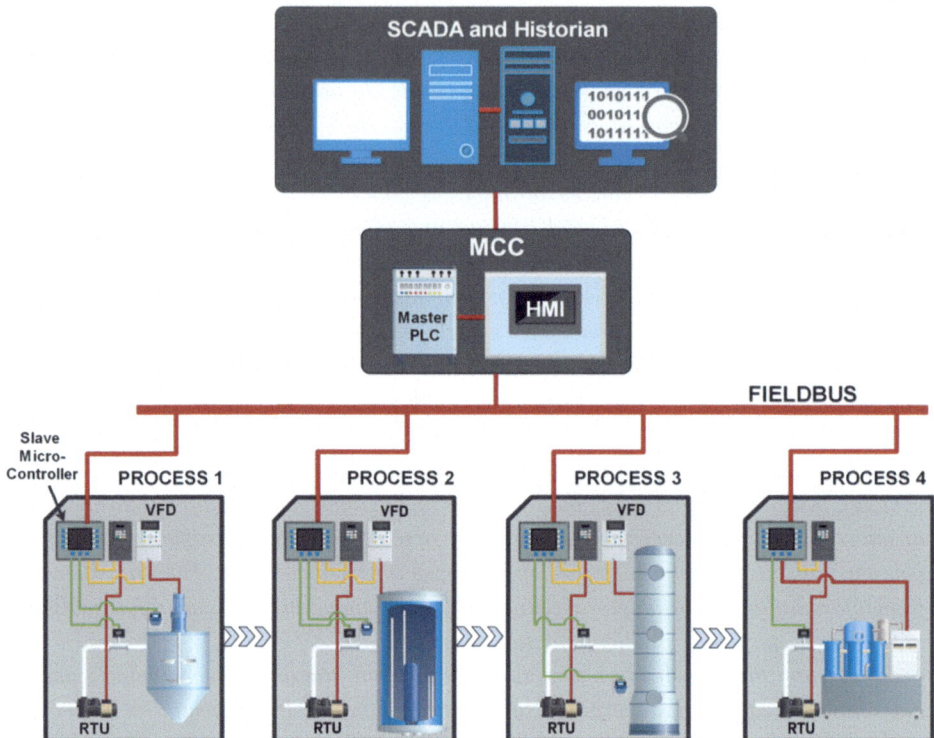

Fig. 4.3 A schematic of the distributed control system (DCS) or plant operational technology (OT) architecture. The communication between the SCADA and MCC layer and the individual RTUs are piped through the fieldbus. The DCS architecture can accommodate different types of network topologies such as ring or branched (tree)

troller. Sensors monitor the physical processes of the plant, and actuators intervene to modify the physical states of equipment (such as opening or closing of a valve) to alter the processes of the plants. The controllers read input information from the sensors, and provide output information to the actuators. The controllers can be programmed to trigger specific responses to the stimuli received through the sensors. In this context, the plant automation system emulates the function of a nervous system. Collectively, these sensory and motor components along with their assigned microcontrollers are termed remote terminal units (RTUs)—see Fig. 4.3. These units form a simple closed loop, controlling individual pieces of machinery and process, and constitute the smallest unit of the DCS.

The communication between the sensors, the controller, and the actuators are conventionally achieved through electrical or pneumatic signals. In electrical transduction, each sensor device contains an embedded programmable electronic converter, which can convert the physical interactions of sensory devices with their environment into calibrated electrical signals. For instance, the temperature measured by a temperature sensor, or tank level

4.4 Components of Digital Transformation

measured by a level gauge can be converted to an electric current of a magnitude between four and twenty milliamperes. Conversely, the electric signals and instructions received by converters in actuators alter their action on the environment. For example, the frequency of an electrical signal in the variable frequency drive (VFD) of a pump motor can be changed to alter the motor speed, and consequently the pump's flow rate.

Individual Remote Terminal Units–sometimes also called Remote Telemetry Unit or Remote Tele-control Unit (RTU) are connected to a single focal point within a plant floor, which may be referred to as the master control unit (MCU). This focal point contains a more general-purpose, usually more powerful, microcontroller that can process signals received from a multitude of RTUs and uses a set of instructions (programs or codes) to operate the connected RTUs in a synergistic manner. These general-purpose microcontrollers are often built using programmable logic controllers (PLC). The PLCs can be programmed, and the code reads the sensory signals as well as the states of different RTUs as inputs and utilizes its programmed logic to send output signals to the actuators of the RTUs to orchestrate the performance of the connected process components and machinery.

When a plant consisting of multiple processes is integrated, the microcontrollers and PLCs of the individual processes are connected to a central control or command center at the plant, often termed as the Master Control Center (MCC). At this MCC, the individual machinery or process equipment PLCs and RTUs are connected to a larger microcontroller, which is sometimes a more powerful PLC. The PLC residing in the MCC acts as the master controller that can receive signals from the individual process equipment PLCs (slaves) and can transmit instructions to them. This configuration is often referred to as the Master-Slave configuration. A more modern terminology for this configuration is the Client-Server configuration where the MCC is the client and the various remote terminal units or MCUs form the servers. The main communication pipeline for information exchange is referred to as the fieldbus. The MCC is the location at a plant where operators and engineers can monitor the overall performance of all the individual processes at the plant and can send instructions to the individual process level PLCs to change the set-points and operating variables. The operator interacts with the plant components through a human machine interface (HMI) located at the MCC, although HMIs can also exist at individual process level control stations (such as monitoring screens, and emergency on/off switches).

On top of the MCC, complex plants often have a supervisory control and data acquisition (SCADA) layer, which accumulates all the process conditions and settings, including time traces of the operational data. The essential components of the SCADA are a SCADA historian, which is a database that collects all the plant data, and the Human Machine Interface (HMI), which allows human operators to interact with the plant. Strictly speaking, HMI is a SCADA component which can be located at any level of the DCS hierarchy (for instance alongside the MCC PLC, or the slave PLCs).

The SCADA historian database is often some form of a relational database (typically some variant of sequential query language, or SQL). At the SCADA layer, operators or engineers can monitor the plant performance and analyze historical data to observe the performance trends or troubleshoot operations. Sophisticated SCADA systems can provide instructions to the Master PLC or individual process PLCs; however, in many process plants, SCADA systems are typically used as monitoring and historizing systems. Generally, engineers monitoring a plant operation at the SCADA level contact the on-floor engineers and operators when they detect any anomaly, and the process is modified at the master or slave PLC level by the operators. Communication between the various PLC layers and the SCADA layer are often achieved through a multitude of communication protocols such as HART, MODBUS, Profibus, CANBUS, Ethernet/IP, OPC/UA, etc. These are industry standard communications protocols that are generally restricted within the plant floor or the OT layer.

In some sense, the OT system at many manufacturing sectors is digitized at markedly advanced levels. Many process industry sectors saw the first wave of digitization manifest on the factory (plant) floor through digital control system (DCS) architectures before the business divisions of such industries were digitized. The SCADA historian has been a mainstay data repository in many industries for decades. Feedback automated control loops have been operating critical machinery in many industrial sectors without any human intervention. In some sense, the OT architecture and even the relational databases of SCADA historians can be considered as legacy components of the automation architecture in a digitalization project. These elements have been a mainstay of industrial automation for decades, are quite robust, and have often been built to stringent standards and specifications of reliability. Finally, one notable feature of the OT architecture is the convergence of the plant operation and automation hierarchy to a single localized point (the MCC and the SCADA control station). While the DCS network at the fieldbus level can have different topologies like a ring or a branched network, it is almost universal in automation integration to implement convergence of all automation activities to a single focal point. This was often a choice exercised in industrial control system design to ensure proper human supervision, control, safety, and security. This single focal point approach often leads to a key bottleneck during planning and implementation of a retrofit digital transformation project. It often creates a limitation in data transfer rates between the OT and the broader enterprise-wide network, and also constitutes a single critical point of vulnerability during a cyberattack.

4.4.2 Information Technology (IT)

Information Technology (IT) serves as the backbone of digital transformation, enabling seamless communication between the physical and digital realms through a combination of hardware, software, and networking technologies. Key digital technologies driving IT advancements include cloud computing, edge computing, artificial intelligence (AI), big data analytics, and cybersecurity frameworks. These technologies allow industries to process vast amounts of data in real-time, facilitating data-driven decision-making, automation, and intelligent control across industrial operations.

4.4 Components of Digital Transformation

One of the most transformative aspects of IT in digital transformation is its ability to enable seamless large-scale communication between plant-floor systems and enterprise-wide digital platforms utilized in the business and corporate realms. Traditional industrial automation systems were confined to local SCADA and Distributed Control Systems (DCS), but IT-driven digital transformation extends these capabilities to enterprise-wide networks and cloud-based platforms. By integrating Industrial IoT (IIoT) devices, smart sensors, and digital twins, IT enables remote monitoring, predictive maintenance, and process optimization, breaking the barriers of conventional, localized automation.

The IT infrastructure comprises both hardware and software elements. Hardware components include edge devices, industrial PCs, data servers, cloud computing infrastructure, and networking hardware such as routers, switches, and industrial gateways. Edge-to-cloud computing architectures allow data to be processed closer to the source (on the edge) before being transmitted to centralized cloud platforms for deeper analytics. This reduces latency, enhances security, and optimizes bandwidth usage. On the software side, IT employs real-time operating systems (RTOS), database management systems, cloud platforms, AI/ML frameworks, and cybersecurity solutions. Communication protocols such as MQTT, OPC/UA, and RESTful APIs facilitate interoperability, ensuring a seamless and secure connection between cyber-physical systems.

Ultimately, IT serves as the key enabler of digital transformation, integrating data-driven intelligence, automation, and connectivity across the industrial landscape. By enabling smart factories, remote diagnostics, and AI-powered decision-making, IT ensures that enterprises remain agile, efficient, and competitive in the era of Industrie 4.0 and beyond. During the four decades leading up to Industrie 4.0, IT massively transformed business operations globally. From enterprise resource planning (ERP) to emails, sharable calendars, and collaborative digital documents, IT has transformed how business processes have become digital, and made operations and decision-making remarkably agile. However, the integration of the factory floor was generally outside the realm of business IT proliferation. Perhaps the biggest design innovation in Industrie 4.0 was the realization of breaking this barrier between IT and OT. As an outcome of this revolution, new integrations between information sharing platforms are evolving, such as the integration of classical MES and CMMS tools with CRM, ERP and supply chain management (SCM) platforms.

4.4.3 The Internet of Things (IoT)

The Internet of Things (IoT) refers to a network of interconnected devices that communicate and exchange data over the internet or private networks, enabling real-time monitoring, automation, and decision-making. IoT devices include sensors, actuators, and embedded systems, all of which generate and share data without requiring direct human intervention. This technology has applications across industries, from smart homes and healthcare to manufacturing and industrial automation.

In the context of digital transformation, IoT plays a pivotal role by enabling industries to collect, analyze, and act on real-time data. Traditional industrial systems relied on manual inspections and periodic data collection, but IoT devices offer continuous, high-resolution monitoring of operational parameters. A key differentiation between IoT and conventional Remote Terminal Units (RTUs) in industrial control systems (ICS) is the scalability, interoperability, and intelligence of IoT devices. RTUs have been used for decades in SCADA systems, collecting process data and transmitting it to a central control station. However, RTUs are typically limited to predefined functions and protocols, whereas IoT devices can operate with greater flexibility, leveraging edge computing, wireless communication, and AI-driven analytics to provide actionable insights in real time. Additionally, IoT enables a distributed intelligence approach, reducing the dependency on centralized control systems and improving resilience and adaptability in industrial automation.

IoT is a key enabler of digitalization, bridging the physical and digital worlds by providing real-time visibility and automation. It facilitates seamless data flow between assets, enterprise systems, and cloud platforms, supporting applications such as predictive maintenance, digital twins, and remote operations. As industries move toward fully autonomous and self-optimizing systems, IoT will continue to be a cornerstone of the next generation of smart manufacturing and industrial automation.

4.4.4 Database as a Single Source of Truth

The primary goal of any digitalization initiative is to aggregate all of an organization's information into a database. A database is a repository of data, a "single source of truth". Databases are of two predominant types, "relational" and "non-relational".

Relational databases are most common in today's industrial practice. Someone familiar with any form of sequential query language (SQL), such as MySQL, POSTGRESQL, or MSSQL, is essentially dealing with relational databases. These databases use tables with pre-defined rows and columns, much like spreadsheets. Data is stored in a structured format with relationships between tables using primary keys and foreign keys. These are best for structured data with strong predefined relationships. It is difficult to alter the structure and organization of a relational database. This is one of the key reasons to make a detailed plan when relational databases are adopted in a digitalization strategy. Any future or incremental expansion of a relational database will be more complex to integrate.

If one is familiar with "MongoDB" or have encountered "JSON", then one is dealing with non-relational databases, also known as noSQL databases. These use flexible data models, such as key-value pairs, documents, graphs, or wide-columns instead of fixed tables. Non-relational databases are more suited for large-scale, unstructured, or rapidly changing data, like social media, IoT, or real-time analytics. Non-relational databases provide more flexibility in terms of expansion, but are often more cumbersome and slower to query or search.

4.4 Components of Digital Transformation

The process of searching and checking out data from a database is achieved through "queries", which are instructions to seek specific data segments from a database, perform some aggregations on those, and transform these into desired representations constituting knowledge. People can query data from an existing database and create new data architectures.

In many cases, large databases need large cloud infrastructure for storage. In modern database terminologies, large databases containing huge amounts of information are given names like data lake and data ocean to signify how vast such databases can be. Every data source in the organization can be connected by appropriate pipes to the data lake. As data flows into the lake from different sources, it is also possible to create outflows of data from the lake through appropriate queries.

4.4.5 Computation for Digital Transformation

The essence of digital transformation lies in the ability of computational systems to continuously shadow physical operations, query data, and provide both insight and foresight about a plant's processes in real time. Unlike traditional computing, which was often reactive, modern digitalized plants leverage an ecosystem of hardware and software that operates round the clock to enable continuous monitoring, predictive analytics, and intelligent decision-making. This shift represents a fundamental transformation in computational philosophy—one that moves from sporadic, operator-driven data analysis to an autonomous, distributed, and intelligent computational fabric that enhances operational efficiency, agility, and resilience.

Edge to Cloud Integration

A key enabler of continuous computation is the distributed nature of computing platforms spanning from edge devices near the plant floor to cloud-based high-performance computing systems. Edge computing devices, such as programmable logic controllers (PLCs), industrial PCs, and IoT gateways, are positioned at the operational technology (OT) layer to collect and preprocess data in real-time. These devices act as the first line of intelligence, performing localized calculations, filtering irrelevant data, and ensuring low-latency responses to critical control operations.

As data moves upward, fog computing nodes, typically found in on-premise industrial servers or data centers, provide intermediate computation, aggregating information from multiple sources and performing deeper analytics. Finally, cloud computing platforms house enterprise-wide datasets, executing complex AI-driven analytics, machine learning (ML) models, and large-scale simulations to generate actionable foresight. The cloud also serves as a repository for historical data, allowing for long-term trend analysis, anomaly detection, and performance optimization across multiple plant sites.

Communication, Computation, and Data Management

For this distributed system to function efficiently, seamless communication protocols are required to ensure low-latency, high-throughput, and secure data exchange. Industrial networks rely on protocols such as MQTT, OPC/UA, and industrial Ethernet standards to facilitate real-time machine-to-machine (M2M) and machine-to-enterprise (M2E) communication. These protocols allow sensor data, operational metrics, and AI-generated insights to flow dynamically between edge, fog, and cloud layers, ensuring that decision-making processes are both timely and data-driven.

On the software side, operating system (OS) selection plays a crucial role in maintaining high system uptime and efficiency. At the edge, real-time operating systems (RTOS) ensure deterministic behavior for mission-critical operations. Mid-tier computing layers often utilize Linux-based OS platforms, given their stability and flexibility in handling industrial workloads, while cloud platforms run on highly scalable environments such as Kubernetes-orchestrated containerized applications for dynamic workload management.

The database management system (DBMS) structure also evolves with digital transformation. Traditional relational databases (SQL-based systems), which were previously dominant in industrial settings, are now complemented by NoSQL and time-series database architectures that allow for efficient storage and querying of high-frequency sensor data. These databases support both structured and unstructured industrial data, enabling advanced analytics and AI-driven insights.

Analytics and Machine Learning in Distributed Computing

A major differentiator of digital transformation is how computation is used not just for insight (reactive analysis) but also for foresight (predictive and prescriptive analytics). Advanced machine learning algorithms, from statistical regressions to deep neural networks, continuously process vast amounts of plant data to detect patterns, anticipate failures, and optimize performance.

At the edge, lightweight AI models and rule-based decision trees can execute real-time anomaly detection, ensuring that equipment operates within safe parameters. At the fog level, predictive maintenance models leverage historical and real-time data to forecast potential failures before they occur, reducing unplanned downtime. At the cloud level, reinforcement learning and prescriptive analytics can simulate plant-wide optimizations, recommending operational strategies that maximize throughput, efficiency, and sustainability.

This distributed approach ensures that the computational load is balanced across different layers, optimizing latency, bandwidth, and cost while ensuring uninterrupted decision support for plant operators and enterprise decision-makers.

From Reactive to Proactive Computing

One of the most significant impacts of digital transformation is the transition from reactive computing (which only responds when issues arise or a programmer triggers the computation) to proactive, autonomous computing that continuously predicts, optimizes, and advises operators before inefficiencies occur.

Prior to digital transformation, industrial computing was largely episodic, where control systems were configured to execute predefined logic without real-time adaptation to changing conditions. Data logging was often performed in batch mode, and insights were derived manually through offline analysis, making it difficult to dynamically adjust operations in response to evolving plant conditions.

With digital transformation, computation is now context-aware, continuous, and predictive. AI-driven algorithms dynamically adjust control parameters, optimize energy consumption, and minimize waste, leading to a fundamental shift in plant management philosophy. Operators are no longer burdened with micromanaging every process variable; instead, they oversee an intelligent computational system that provides recommendations and automated corrective actions, drastically improving efficiency and reliability.

Computation is the lifeline of digital transformation, enabling an industrial plant to function as a cyber-physical system that continuously shadows, analyzes, and optimizes real-world operations. Through an orchestrated combination of edge, fog, and cloud computing, a modern digitalized plant can operate with autonomous decision-making capabilities, ensuring that data-driven insights translate into real-world business value.

This shift is not just about efficiency gains, it fundamentally redefines how industrial plants operate, adapt, and evolve. The ability to maintain round-the-clock computation, seamlessly integrate across distributed networks, and leverage AI-driven foresight marks the key difference between pre-digitalization reactive operations and post-digitalization proactive intelligence. Digital transformation, therefore, is not just about digitizing data—it is about computationally reimagining how industries function.

4.5 Closing Remarks

This chapter introduces fundamental technical terminology and provides a preview of key concepts that will be explored throughout this book. At the core of digital transformation lies the DIKW framework—a structured approach that moves from raw Data to Information, then to Knowledge, and ultimately to Wisdom. The primary objective of digital transformation is to enhance the value derived from data, enabling organizations to transition from mere data collection to actionable insights. However, wisdom for its own sake holds little value; instead, digitalization efforts must be guided by a clear purpose. The ultimate goal should be to leverage data-driven wisdom to improve operational efficiency, generate business value, enhance agility, and build resilience in industrial processes.

A key technical pillar of digital transformation is the seamless integration of Operational Technology (OT) and Information Technology (IT). OT encompasses the sensors, control systems, and industrial automation technologies that manage plant operations, while IT consists of enterprise-level systems that handle data processing, analytics, and decision-making. Successful digital transformation hinges on bridging these traditionally separate domains, enabling a unified ecosystem where operational data flows seamlessly into business intelligence systems.

Beyond integration, the effectiveness of digital transformation depends on the ability to extract meaningful insights and foresight from the integrated data infrastructure. A well-structured digitalization initiative must rationalize how additional value can be unlocked through IT-OT convergence and how the combined dataset can be efficiently analyzed in real time through a well orchestrated computational infrastructure. This includes leveraging advanced analytics, AI-driven decision support systems, and predictive modeling on a distributed and round the clock operating computational network to drive smarter, data-informed strategies. By architecting a digital ecosystem that prioritizes both real-time process insights and long-term strategic foresight, organizations can unlock new opportunities for optimization, innovation, and sustained growth.

References

1. Henning Kagermann, Wolfgang Wahlster, and Johannes Helbig. Recommendations for implementing the strategic initiative industrie 4.0: Final report of the industrie 4.0 working group. Technical report, acatech, Munich, Germany, 2013.
2. Russell L. Ackoff. From data to wisdom. *Journal of Applied Systems Analysis*, 16:3–9, 1989.
3. Keith A. Stouffer, Victoria Y. Pillitteri, Suzanne Lightman, Marshall Abrams, and Adam Hahn. Guide to industrial control systems (ics) security. Special Publication (NIST SP) 800-82 Rev 2, National Institute of Standards and Technology, 2015.

Part II
The Journey Toward Digital Transformation

Planning Digital Transformation 5

5.1 Implementing Digitalization—The Roadmap

The journey toward digital transformation should begin with a well-structured concept development and planning phase before moving into implementation. Unlike conventional new product or process introduction (NPI) procedures commonly used in the process industry, digital transformation presents unique challenges. Traditional NPI processes often follow a stage-gated approach or a technology readiness level (TRL)-based evaluation scheme, where progress is assessed step by step. While such structured methodologies can be applied to digitalization, it is critical to recognize that proof-of-concept efforts cannot be conducted in isolation or rely solely on historical data slices.

Approaches that attempt to validate digital transformation in a controlled "laboratory-test" setting often fail to capture the complexities of real-world implementation. This disconnect can lead to misleading results and a wasteful scaling effort that does not translate into practical benefits. Instead, digitalization projects are more likely to succeed when they are designed from the outset to operate within the actual process plant environment, addressing its technical, operational, and organizational intricacies. By grounding digital transformation initiatives in real-world conditions and aligning them with plant dynamics, companies can unlock tangible results, maximize efficiency, and drive meaningful improvements across operations.

In this chapter, we will highlight the critical considerations for proponents of digital transformation, focusing on concept development and planning. Specifically, we will examine the organization's pre-digitalization state, including its manufacturing environment and operational processes. We will then explore how these processes will be modified through digital transformation and define the desired end-state of this transformation journey. Given the complexity of this undertaking, which requires multidisciplinary expertise and cross-functional collaboration, securing top-down buy-in from leadership will be essential for its success.

It is important at the outset to have clarity about the magnitude of effort required for digital transformation, primarily by addressing the following three main questions:

1. *Is digital transformation a worthwhile endeavor for the plant or enterprise under consideration?* In many cases, a plant, manufacturing process, or enterprise may already be highly optimized, making the incremental gains from an additional digital transformation effort negligible. Conversely, certain enterprises or manufacturing operations rely on legacy processes that are difficult to automate, generate minimal or highly abstract data, or operate in niche sectors requiring extensive modifications to both their processes and digital infrastructure. In such cases, the economic justification for digital transformation must be carefully evaluated.
2. *What are the requirements for a successful transition?* A well-structured transition demands a comprehensive planning phase that assesses the current state of the manufacturing plant, the underlying processes, the already existing degree of digitization of information, and the resources, including financial, technical, and human, required to facilitate the transition. Properly allocating time and effort in this stage ensures a smoother and more effective implementation.
3. *What tangible benefits will the enterprise derive from digital transformation?* Defining clear and measurable outcomes is critical when setting goals and aspirations. These should go beyond generic metrics such as improved efficiency, increased profitability, or enhanced sustainability. Instead, they should be enterprise-specific and quantifiable. For example: (i) achieving a fourfold acceleration in the production cycle from raw material intake to finished product packaging, or (ii) reducing the frequency of unexpected plant shutdowns due to unscheduled maintenance by 20%. Setting realistic, context-driven expectations while maintaining some flexibility in the final metric to be achieved is key to ensuring the success of the digital transformation initiative.

Digital transformation in manufacturing is fundamentally the automation of workflows through data-driven methodologies, relying on the seamless integration of digital technologies into industrial processes. This transformation makes IT frameworks, distributed computational architectures, and extensive databases central to operations, enabling real-time data collection, analysis, and decision-making. By leveraging these technologies, organizations can optimize processes, enhance efficiency, and improve predictive maintenance, ultimately driving higher productivity and operational agility.

However, many process industries may not have the necessary infrastructure or in-house expertise to support such an extensive digital transformation. Legacy systems, limited IT capabilities, and a lack of experience in data-driven decision-making can pose significant barriers. Therefore, rather than adopting a generic, top-down approach to digitalization, organizations must develop transformation strategies that align with their specific operational needs. It should be borne in mind that digital transformation is not merely acquisition of IT, automation, and computational assets, but a systematic change in how the humans in an enterprise interact and engage in workflows. A successful digital transformation initiative should begin with a comprehensive assessment of existing workflows, automation levels, and key challenges. By focusing on targeted, value-driven implementation rather than broad, indiscriminate digitalization, enterprises can ensure that the integration of digital technologies leads to greater efficiency, streamlined operations, and tangible business benefits.

5.2 Concept Development and Planning

The initial step of any digital transformation project is a careful assessment of the baseline state of the plant, business organization, operational practices, and procedures. The assessment phase of a digital transformation journey consists of the following key tasks:

1. Evaluating the current state of digitization within the organization.
2. Establishing the team tasked with planning a implementation.
3. Ensuring seamless communication between the plant floor and the business operations through IT/OT integration.
4. Assessing the data structures, information processing, and automation algorithms needed to enhance organizational and product workflows into agile systems.
5. Assessment of the desired interaction of the digital infrastructure with the plant and business operations teams, focusing on the desired user experience.

In Chaps. 2 and 3, we provided the business, technical, and strategic rationale for the pursuit of digitalization. Beyond these technical and strategic assessments, an essential guiding principle must underpin any digitalization initiative: it should enhance human roles, improve quality of life, and foster greater engagement and service efficiency. The success of digital transformation depends on the active participation and support of those who operate and manage industrial processes.

A critical lesson from several early factory digitalization initiatives is that many efforts either failed or yielded limited benefits. In most cases, these shortcomings arose because stakeholders did not perceive measurable improvements aligned with business case expectations. In other instances, digitalization efforts introduced overwhelming data management workloads or created a sense of uncertainty or threat among plant operators and management personnel. As a result, many of these initiatives were ultimately rejected or abandoned.

For digital transformation to succeed, human adoption is paramount. Without the buy-in and engagement of the workforce, even the most advanced digital initiatives risk being deemed ineffective or unsustainable.

5.2.1 Planning Digital Transformation for an Enterprise

Digital transformation is a multi-phase arduous process that integrates plant floors to the highest echelons of the enterprise to create a connected and data-driven enterprise. A successful transformation requires careful planning, execution, and ongoing improvements to maintain efficiency and security.

During an initial strategic planning and vision development phase, the digital transformation team should be tasked to create a strategic vision and a roadmap that should be acceptable to the entire organization. This blueprint document should minimally include

- **Defining Business Objectives**: Establishing transformation goals aligned with operational efficiency, customer experience, or predictive maintenance specific to the enterprise. For example, a manufacturing plant may seek to integrate real-time process monitoring with advanced analytics to enhance production quality and reduce waste.
- **Assessing Current Digital Maturity**: Evaluating legacy IT/OT infrastructure, data silos, workforce digital competency, and existing cybersecurity frameworks.
- **Identifying Key Stakeholders**: Engaging leadership teams, department heads, IT personnel, and OT engineers to ensure cross-functional collaboration.
- **Establishing Key Performance Indicators (KPIs)**: Measuring success through improved production efficiency, reduced downtime, enhanced decision-making, and optimized resource utilization.

It is important to ensure that the digital transformation strategy and roadmap is approved by management, and the ensuing changes it will create in the workflow and plant environment is agreed upon by the employees at all levels of the enterprise. The digital transformation team should spend time to disseminate and discuss the strategic plan with all stakeholders, and even refine or revise this plan to ensure it ha buy in from the entire organization.

5.2.2 Key Considerations for Digital Transformation Success

When planning digital transformation, organizations must ensure that the plan is not overambitious, is relevant, and is affordable. The long-term sustainability of the initiative is of critical importance. Therefore attention to the following factors at the outset is important.

Technology Readiness: Legacy systems must be assessed for compatibility with modern digital infrastructure. For example, older control systems using proprietary protocols may need gateways or middleware to interface with modern industrial networks. Business process that are not digitized also need to be upgraded.

Data Utilization and Analytics: The transformation strategy must define how data will be collected, processed, and utilized. This includes considerations for real-time analytics, historical trend analysis, and predictive modeling.

5.2 Concept Development and Planning

Cybersecurity and Compliance: Data security must be a priority to protect against cyber threats. Role-based access control, encryption, and continuous monitoring should be implemented to secure both IT and OT networks.

Scalability and Flexibility: The chosen architecture should allow for modular expansion. For instance, cloud-based platforms provide scalable storage and processing power to accommodate growing data volumes.

User Adoption and Workforce Training: Successful implementation depends on user adoption. Training programs should be tailored to different roles, ensuring operators, engineers, and analysts can effectively use the new digital tools.

Digital Infrastructure as a Continuous Asset: A digital transformation strategy must treat the developed information pathways, digital hardware and software infrastructure as continuously evolving assets requiring ongoing maintenance. This includes putting in place procedures for

- Regular maintenance and upgrades of the industrial control systems, data processing nodes, and cloud services to maintain performance and security.
- Data governance and quality assurance with systematic data validation methods to ensure accurate insights and decision-making.
- Software and firmware updates to optimize algorithms for data processing and automation and reduce vulnerabilities.
- Disaster recovery and business continuity plans such as backups and redundancy mechanisms to ensure minimal disruption during failures.

5.2.3 Tools for Planning and Execution

Digital transformation teams can leverage various platforms and methodologies for implementation, such as:

Business and Industrial Process Modeling: Used to map out digital workflows, automation, and process optimization pathways.

Project Management Methodologies: Agile frameworks enable iterative deployment and stakeholder feedback integration.

IT/OT Integration Protocols: Open-standard protocols facilitate seamless data exchange between industrial devices and enterprise IT systems.

Data Analytics and AI Platforms: Predictive models enhance operational intelligence and enable condition-based maintenance strategies.

Cybersecurity Solutions: Secure network segmentation and anomaly detection systems safeguard industrial assets.

During the planning and strategy development phase of an enterprise digital transformation process, a collaborative Knowledge Management System (KMS) hosted on a cloud based platform plays a pivotal role in ensuring seamless information sharing, documentation, and strategic alignment across teams. A well-structured KMS allows stakeholders from various departments, such as operations, IT, engineering, finance, and management, to contribute their insights, ensuring that the digital transformation vision is comprehensive and reflects the diverse needs of the organization. By integrating structured data (such as technical documents, process workflows, and implementation roadmaps) with unstructured knowledge (such as expert insights, discussions, and lessons learned), the KMS fosters a dynamic, evolving blueprint for transformation. As the digital transformation progresses, this repository becomes the central single source of truth for the enterprise, storing records of technology adoption, best practices, system integrations, and performance benchmarks. Over time, the KMS evolves beyond a planning tool into an enterprise-wide knowledge hub, supporting continuous learning, workforce training, and adaptive decision-making. By incorporating features such as version control, role-based access, and real-time collaboration, a well-implemented KMS ensures that the organization maintains institutional memory and sustains long-term digital evolution in an agile and structured manner.

> *Knowledge Management Systems (KMS)*
> *A Knowledge Management System (KMS) is a digital platform designed to collect, store, organize, and share an organization's knowledge resources. KMS facilitates efficient knowledge retrieval, collaboration, and decision-making by providing structured access to both explicit knowledge (documents, manuals, databases) and tacit knowledge (expert insights, workflows, best practices). These systems are widely used in enterprises for digital transformation, research, customer support, and employee training.*
>
> *Examples of Knowledge Management Systems*
> *Wikipedia is the best example of a knowledge management system we all are familiar with. In enterprise environments, there are several KMS solutions available, including Confluence—A collaborative documentation and knowledge-sharing platform, SharePoint—A Microsoft-based content management system with enterprise integration, MediaWiki—A wiki-based knowledge management system, used by Wikipedia and enterprises, and Alfresco—A content services platform with knowledge repository features.*
>   *Several open-source KMS platforms offer robust capabilities and flexibility for enterprises, such as, Wiki.js—A lightweight, modular wiki platform that supports Markdown and database integrations, DocuWiki—A simple file-based wiki designed for documentation, OpenKM—A document management system with workflow automation and knowledge indexing, and XWiki—A powerful structured wiki that supports enterprise collaboration and document organization.*

> **Hosting and Maintenance of KMS**
> KMS platforms can be self-hosted on on-premise servers or cloud-hosted. Self-hosted solutions require dedicated infrastructure, while cloud-hosted options offer scalability and remote accessibility. Maintenance efforts typically include regular updates & security patches, backup & data integrity management, user access management—controlling permissions and role-based access to ensure knowledge security, integration with other enterprise systems, such as, linking KMS with enterprise resource planning (ERP), document management systems, and digital transformation tools, as well as, content curation & quality assurance—Ensuring knowledge remains relevant, structured, and searchable.
>
> A well-maintained KMS becomes a living knowledge repository, supporting continuous improvement, onboarding, collaboration, and the long-term success of digital transformation initiatives.

5.2.4 Workforce Training and Change Management

The success of digital transformation hinges on workforce readiness. Organizations must plan for allocating time and resources to train the workforce to adopt the new workflows, standard operating procedures, and productivity tools. This requires,

- Development of structured training programs tailored to new technologies.
- Promoting up-skilling initiatives, such as certification programs in industrial automation and cloud computing.
- Implementing change management frameworks to reduce resistance and ensure smooth transition.
- Establishing knowledge-sharing portals and help desks for continuous learning.

5.2.5 Timeline and Continuous Improvement Strategy

A phased approach ensures gradual and manageable transformation. Expected timelines for a digital transformation initiative are:

- **Short-Term (0–6 Months)**: Conduct assessment, define strategy, select technologies, and implement pilot projects.
- **Mid-Term (6–18 Months)**: Scale implementation, refine processes, train employees, and optimize cybersecurity.
- **Long-Term (18+ Months)**: Continuously evaluate, upgrade, and integrate emerging technologies such as AI and digital twins.

A structured transformation roadmap ensures that technology adoption remains sustainable and aligned with business objectives. Successful digital transformation requires strategic vision, robust IT/OT integration, and a continuous improvement mindset. Organizations must prioritize cybersecurity, workforce training, and scalable architectures to ensure long-term success. By leveraging advanced analytics, cloud computing, and Industrial IoT, businesses can unlock new efficiencies, reduce operational costs, and enhance decision-making capabilities.

5.3 Digitization: Foundation of Digitalization

A digital transformation journey consists of two steps —digitization followed by digitalization. Digitization refers to the process of converting or encoding analog data (such as manually recorded information) into a digital format. It represents the transition from analog to digital record-keeping. This step becomes necessary if the process plant environment still relies on manual (paper-based) data collection, maintenance record-keeping, inventory management, and other similar processes. While several manufacturing sectors have advanced significantly in data integration and analytics, plant operations in many areas of the process industry remain insufficiently digitized for the effective execution of digital transformation initiatives. Additionally, if such data exist in fragmented or disconnected repositories (databases), integration into a centralized data lake or data ocean as a single source of truth is essential.

Operational data are typically collected by a plant's automation and process control systems. Basic plant instrumentation data are recorded in the SCADA historian, which structures information as time-series data, capturing process variables at predefined intervals. Even if certain data points are not initially recorded by the historian, they can often be added with minimal effort. Upgrading the SCADA historian to be readily accessible to all relevant stakeholders across the organization has proven to be a highly beneficial enhancement. Unlocking access to the historian beyond the plant floor enables organizations to measure plant performance effectively and integrate operational insights with broader business databases and performance monitoring tools.

The existing digitization frameworks within an organization should be carefully evaluated. In addition to the SCADA historian, various other plant and process performance data management systems may already be in place, such as inventory management systems, maintenance systems, and enterprise information management platforms. Table 5.1 lists various types of databases that can be commonly associated with manufacturing environments. The most commonly used systems include the Manufacturing Execution System (MES), Computerized Maintenance Management System (CMMS), Laboratory Information Management System (LIMS), and Enterprise Resource Planning (ERP) systems. If these systems are already embedded within the organization, the necessary infrastructure for digitizing operational and process data likely exists. These systems also facilitate the

5.3 Digitization: Foundation of Digitalization

Table 5.1 A list of database systems and repositories used in a plant digital transformation scope

Name	Abbreviation	Purpose	Data organization	Key data content
Asset management database	–	Tracks plant assets, equipment lifecycle, and maintenance history	Relational (SQL-based)	Equipment lists, bill of materials (BOM), maintenance logs, asset tags
Manufacturing execution system	MES	Manages real-time production execution, workflow tracking, and resource allocation	Hierarchical, event-driven	Work orders, production schedules, machine utilization, real-time shop floor data
Computerized maintenance management system	CMMS	Automates maintenance workflows, schedules repairs, and manages spare parts inventory	Relational (SQL-based)	Preventive maintenance schedules, work orders, failure reports, maintenance logs
Laboratory information management system	LIMS	Manages laboratory sample tracking, test results, and quality compliance	Relational, document-based	Sample IDs, test results, chemical compositions, regulatory compliance reports
SCADA historian database	–	Stores time-series process control and automation data for operational analysis	Time-series database	Temperature, pressure, flow rates, alarms, event logs, real-time telemetry data
Customer relationship management	CRM	Tracks customer interactions, sales activities, and service records	Relational (SQL-based), NoSQL for big data	Customer details, sales pipeline, order history, support tickets, marketing campaigns
Enterprise resource planning	ERP	Integrates business functions like finance, HR, supply chain, and procurement	Relational (SQL-based), distributed systems	Financial transactions, procurement data, employee information, inventory records
Product lifecycle management	PLM	Manages product design, engineering changes, and compliance tracking	Object-oriented	CAD files, BOM, revision history, compliance documents, supplier details
Supervisory control and data acquisition	SCADA	Provides real-time process monitoring, automation, and remote control of equipment	Distributed, event-based	Sensor data, process control parameters, alarms, control setpoints, operational logs
Digital twin database	–	Represents a virtual model of physical assets for simulation and predictive analysis	Graph database, real-time event processing	Sensor readings, asset parameters, simulation models, historical performance data
Industrial Internet of Things	IIoT	Collects and integrates data from connected devices and smart sensors for analytics	NoSQL, time-series database	IoT sensor data, real-time monitoring logs, predictive maintenance insights
Human-machine interface	HMI	Provides user interfaces for operators to interact with control systems and visualize process data	Distributed, event-driven	Operator commands, alarms, real-time equipment status, process dashboards
Quality management system	QMS	Ensures compliance with industry standards and manages quality assurance processes	Relational, document-based	Non-conformance reports, corrective actions, quality inspection records, regulatory compliance data

integration of production-related information with critical aspects of manufacturing operations, including supply chain management (SCM) and customer relationship management (CRM) databases.

For example, the presence of a CMMS typically ensures a structured approach to plant maintenance scheduling. Such a system provides access to databases containing operational data on each piece of equipment, maintenance schedules, historical repair records, and maintenance cost tracking. Leveraging this information enables more efficient asset management and supports data-driven decision-making in maintenance operations.

If digitization has already been implemented within the plant, the focus can shift directly to developing a roadmap for digitalization goals and implementation steps.

Digitalization involves utilizing digitized data to generate insights, optimize processes, and achieve organizational goals. It represents the core focus of the digital transformation initiatives, where data-driven strategies will drive efficiency and innovation. A critical step in this process is identifying which data hold value and understanding the benefits of storing, analyzing, and processing such data.

In context of digitization of a chemical manufacturing environment, initiatives are required in ensuring that all pertinent measurements of the physical plant conditions are stored in a consistent digital format that accurately represents the state and time history of the plant and the processes. Furthermore, such time series information needs to be mapped against the basic design and equipment specifications of the plant. The approach taken depends on a careful assessment of the state of an existing plant (in case of retrofit digitalization), or needs to be defined solidly during the design phase of the plant for greenfield projects. Noting that older generation plants were not designed with a "digital first" mindset, considerable effort needs to be devoted in retrofit digitalization projects to ensure that the relevant measures of physical parameters are digitized properly, such that they can provide a meaningful analysis of the performance of the plant with adequate resolution and accuracy. This is a formidable task in itself, often limiting the outcomes of many digitalization projects.

Digitization in chemical manufacturing enterprises requires the systematic conversion of critical operational data into structured digital formats. The goal is to ensure seamless integration, retrieval, and analysis for process optimization, regulatory compliance, and decision-making. Below, we outline the key information categories that need digitization, specifying their storage formats, data structures, and integration requirements.

5.3.1 Key Information for Digitization

The key information about a process plant that needs to be considered for digitization includes:

1. **Process Design Documents**: Essential process plant design elements should be digitized, particularly:

5.3 Digitization: Foundation of Digitalization

- **Process Flow Diagram (PFD)**
- **Piping and Instrumentation Diagram (P&ID)**
- **Electrical and Control Schematics**, such as single line diagrams (SLD)
- **Network Architecture Diagram**
- **Control Logic Representations**, such as functional block diagrams (FBD) and instrument loop diagrams.

These documents should be in digital formats, with machine-readable labels or tags to facilitate integration with other digital systems.

2. **Bill of Materials (BOM)**: A detailed inventory of plant equipment, including:

   - Equipment specifications and properties
   - Mapping of equipment to PFD and P&ID labels or tags
   - Asset mapping and cataloging using database tools

   A well-structured BOM enables efficient asset tracking, maintenance scheduling, and digital twin integration.

3. **User and Operating Manuals**: Standard operating procedures (SOPs), user manuals, and maintenance guides should be digitized. If available in print, converting them into machine-readable formats allows for:

   - Efficient retrieval and reference in digital workflows
   - Utilization in AI-driven knowledge systems such as large language models (LLMs) for training and troubleshooting

4. **Operational and Control Philosophy**: A structured repository should document:

   - Interlocks and safety mechanisms
   - Control loop descriptions
   - Overrides and fail-safe operations

   This information is crucial for automation logic validation and regulatory compliance.

5. **Process Control Program Listings**: Digital storage of control system configurations, particularly:

   - Remote Terminal Units (RTUs), Programmable Logic Controllers (PLCs), and Supervisory Control and Data Acquisition (SCADA) system programs
   - Process tags with standardized naming conventions mapped across PFD, P&ID, SLD, and control logic
   - Classification of tags into:

- Analog Input (AI)–Sensor data
- Analog Output (AO)–Control instructions
- Digital Input (DI)–Binary status signals
- Digital Output (DO)–Actuator control signals

Proper cataloging of these data points ensures robust automation and analytics capabilities.

6. **SCADA Historian Data**: Continuous and time-series data from the plant's SCADA system should be stored in historian databases, capturing:
   - Real-time process variables such as temperature, pressure, flow rates, and energy consumption
   - Long-term trend data for predictive analytics and anomaly detection
   - Structured time-stamped datasets for integration with AI-based optimization models

7. **Laboratory Information Management System (LIMS) Data**: Chemical and quality control analysis data from laboratory systems should be digitized, including:
   - Raw material and product quality measurements
   - Batch tracking and sample identification
   - Compliance and regulatory reporting datasets

Standardized LIMS data structures enable advanced analytics, statistical process control, and seamless integration with production data.

Digitized Data:

The digital I/O tags are binary bits with only two states (On and Off), whereas the analog input or output (AI/AO) tags are signals that are stored internally as words spanning 2, 4, or 8 bytes, where one byte comprises 8 bits. Therefore, a 2 byte word consists of 16 bits.

Digital inputs and outputs (DI/DO) are bitwise information that indicate state values of real processes (True or False, On or Off). Analog information is stored in digitized form as unsigned (positive) or signed (for both negative and positive) integers. In an unsigned 16 bit storage format, we can store integers ranging from 0 to 65,535 in a word consisting of 2 bytes, whereas in a signed 16 bit format, we can store integer values ranging from −32768 to 32767 in a word.

A process controller (PLC) maps analog data collected from sensors to 16 bit digital representation of integers. For instance, temperatures measured by a temperature sensor with a range of 0 to 200 °C in the physical realm can be mapped to an unsigned integer scale ranging from 0 to 65,535. The real value, the corresponding integer, and the binary representation of the temperature scale are shown in the table below.

5.3 Digitization: Foundation of Digitalization

$Temperature(°C)$	$Unsigned\ Integer$	$Binary\ Value$
0	0	0000 0000 0000 0000
50	16, 384	0100 0000 0000 0000
100	32, 768	1000 0000 0000 0000
150	49, 152	1100 0000 0000 0000
200	65, 535	1111 1111 1111 1111

Note that the mapping between the temperature and the integer is linear and continuous. However, the digital scale has a resolution of

$$Resolution = \frac{(200 - 0)}{2^{16} - 1} = 0.00305$$

In other words, the integer numbers in the digitized scale can only resolve the physical temperature scale with a least-count increment of 0.00305 °C.

5.3.2 Data Structure Coordination Across Digital Platforms

To effectively manage and utilize these digitized datasets, a structured approach is required to coordinate data storage across different digital platforms:

- **Relational Databases (SQL-based)**—Best suited for structured data such as BOM, process control program listings, and operational manuals.
- **Time-Series Databases (Historian DBs)**—Used for SCADA historian data and real-time sensor logs, enabling efficient trend analysis.
- **Document Databases (NoSQL-based)**—Ideal for unstructured data like digitized manuals, SOPs, and LIMS reports.
- **Cloud and Edge Computing Platforms**—Used for real-time processing, AI-driven analytics, and machine learning applications that require synchronized access to multiple data sources.
- **Data Lake Architectures**—Integrate structured, semi-structured, and unstructured datasets from various sources into a unified digital framework.

By ensuring seamless interoperability between these database platforms, chemical manufacturing enterprises can achieve a comprehensive digital transformation, enabling smarter decision-making, predictive maintenance, and operational efficiency.

5.3.3 Digitized Data Quality in Process Automation

The quality of digitized data in industrial automation and process control is crucial for effective decision-making, predictive maintenance, and process optimization. High-quality data ensures reliability, accuracy, and meaningful analysis, while poor data can lead to incorrect inferences, faulty alarms, and inefficient process control strategies. Various challenges, such as sensor noise, incorrect timestamps, inconsistent sampling rates, and improper interpolation methods, can degrade data quality. In process industries where SCADA (Supervisory Control and Data Acquisition) systems are used to log multiple process variables, ensuring high-fidelity time-series data is critical to avoid misleading correlations and erroneous conclusions [1].

Data Quality, Noise, and Statistical Indicators

One of the key challenges in digitized process data is the presence of noise, which can arise due to sensor inaccuracies, environmental interference, and electronic drift. High-quality data should have minimal noise and an appropriate signal-to-noise ratio (SNR) [2]. Noise filtering techniques, such as moving average smoothing or Kalman filtering, can be applied to reduce fluctuations while preserving essential data trends [3]. Table 5.2 summarizes some key quality indicators for time-series data.

To ensure good data quality, historical records should be validated through statistical methods such as mean, standard deviation, skewness, and kurtosis. Proper classification of anomalies—whether they are actual process disturbances or sensor malfunctions—requires statistical validation, machine learning anomaly detection, or physical verification.

Table 5.2 Key quality metrics for time-series data

Quality metric	Description
Signal-to-Noise Ratio (SNR)	Ratio of signal strength to noise level, ensuring reliability
Data completeness	Percentage of missing or invalid records in a dataset
Temporal consistency	Correct sequencing and ordering of time-stamped data
Redundancy checks	Identification of duplicate or inconsistent records
Statistical distribution	Normality and outlier detection through statistical analysis

Accuracy of Time Stamping in Time-Series Data

In SCADA systems, multiple process variables (e.g., temperature, pressure, flow rate) are logged asynchronously, but often they are assigned common timestamps when stored. This can lead to misleading inferences. For example, if a temperature sensor updates every 1 second but a pressure sensor updates every 10 seconds, their data points should not be assigned the same timestamp unless explicitly synchronized. To mitigate this issue, proper time synchronization protocols such as Network Time Protocol (NTP) should be implemented, and interpolation methods should be carefully chosen when aligning different time-stamped datasets.

A common problem arises when process variables with different update frequencies are forced into a uniform timestamp. Suppose a flow sensor records data every 0.5 seconds, while a pressure transmitter logs every 5 seconds, and both are assigned a common timestamp in a SCADA system. This can introduce false correlations, leading to incorrect cause-and-effect conclusions. A better approach is to store data with actual timestamps and apply interpolation techniques when necessary.

Importance of Time Step Size and Data Collection Frequency

The choice of data collection frequency is essential to accurately capture time-dependent dynamics in a process. According to the Nyquist theorem, the sampling frequency should be at least twice the highest frequency component present in the signal. In industrial process monitoring, slow-varying parameters such as tank level or room temperature can be recorded at a lower frequency (e.g., once per minute), whereas fast-changing signals like pressure pulsations or turbine speed should be captured at a much higher rate.

For example, if a vibration sensor monitoring a rotating pump has a critical frequency component at 50 Hz, then the data collection rate should be at least 100 Hz to prevent aliasing errors. If sampled at only 10 Hz, significant oscillatory behavior in the pump would be lost, potentially missing early warning signs of mechanical failure. Thus, determining the optimal sampling frequency requires understanding the underlying physics of the process variable being measured.

Ensuring high-quality, noise-free, well-timestamped, and optimally sampled data is vital for digital transformation efforts in process automation. Proper data validation techniques, statistical checks, synchronization protocols, and Nyquist-based sampling strategies help in making reliable data-driven decisions. By maintaining a structured approach to data acquisition and validation, industries can leverage their digital infrastructure for more effective process control, predictive maintenance, and long-term operational efficiency.

5.4 The Digitalization Implementation Team

During the assessment phase, all individuals and departments responsible for digital transformation must be identified to form a dedicated digital transformation team. This team is tasked with implementing a centralized digitalization effort, a crucial step in ensuring the success of the transformation process.

The Chief Information Officer (CIO) or Chief Digital Officer (CDO) typically leads the organization's efforts to establish a digitization and digitalization infrastructure. Under their leadership, the IT and OT teams develop, maintain, and provide access to the entire digital infrastructure. Their role includes facilitating access to networking and computational resources, managing and integrating databases, integrating OT with IT, and addressing cybersecurity concerns. However, it is equally important to identify internal champions and data users who can effectively utilize digitized data and extract value. Engaging process experts with direct experience in plant operations and core manufacturing ensures that digital transformation efforts align with practical, operational needs.

Additionally, the plant operations team, as the primary beneficiaries of digital transformation, must not feel alienated or threatened by the introduction of digital technologies on the plant floor. Instead, efforts should be made to ensure their active participation. Capturing user experience (UX), anticipating workflow changes, and providing early retraining are essential steps in fostering acceptance and effective utilization of new digital tools. As the last line of defense in plant operations, floor operators rely on decision support systems to enhance their effectiveness, making their involvement in the digital transformation process indispensable.

Early communication and collaboration between IT and OT divisions within the organization are critical for ensuring smooth data access, data flow management, and integration of data reservoirs. A common misconception is that OT and IT barriers are insurmountable, creating a significant challenge in merging factory data with business data for organization-wide digital transformation. Past experiences in the chemical processing industry have demonstrated that overcoming these barriers is essential for the successful implementation of digital transformation initiatives.

In Tables 5.3 and 5.4 some representative key players and planners likely to be included in a digitalization task force in a petroleum production (upstream) facility and a food and beverage manufacturing plant are listed. This is not exhaustive, and in some cases, a single person may fulfill multiple tasks in a team. Also, there are instances when some of the task force members can be picked as external delegates assigned to the team or recruited as consultants.

5.5 OT/IT Integration Assessment

Table 5.3 Typical digitalization task force team in a upstream & midstream oil & gas production company

Personnel position	Typical tasks	Involvement timeline
Chief digital officer (CDO) / Chief Information Officer (CIO)	Leads digital strategy, ensures alignment with business goals, oversees cybersecurity & data integration	Initial Phase, Ongoing
Digital transformation lead	Manages the implementation roadmap, coordinates with IT, OT, and business units	Initial Phase, Ongoing
Process control engineer	Integrates automation systems (SCADA, DCS), ensures real-time monitoring, and optimizes control logic	Initial Phase, Ongoing
Data scientist/AI Specialist	Develops predictive maintenance models, anomaly detection, and data analytics for asset performance	Ongoing, Full-Scale Implementation
IT-OT integration specialist	Connects OT with IT, ensures secure communication between field devices and enterprise systems	Initial Phase, Ongoing
SCADA & historian engineer	Configures and manages real-time data logging, historian setup, and dashboard visualization	Initial Phase, Ongoing
Cybersecurity expert	Establishes secure digital communication channels, implements threat detection, and ensures compliance	Initial Phase, Ongoing
Cloud & Edge computing architect	Develops cloud storage & edge computing solutions for real-time analytics at remote production sites	Ongoing, Full-Scale Implementation
Drilling & Production engineers	Provides domain expertise to ensure digital tools align with production optimization goals	Ongoing
Asset integrity specialist	Uses digital tools for pipeline monitoring, corrosion detection, and equipment reliability analysis	Ongoing, Full-Scale Implementation
Supply chain & logistics expert	Implements digital tracking for supply chain, optimizes procurement and inventory using AI tools	Full-Scale Implementation
Regulatory & compliance officer	Ensures all digital transformation initiatives meet environmental and safety regulations	Ongoing

5.5 OT/IT Integration Assessment

A key focus of any digital transformation assessment initiative is evaluating the current status of operations technology and information technology (OT/IT) connectivity and implementation within an organization, and, if they are disconnected, determining how to bridge these gaps effectively.

To achieve this, it is essential to develop a clear understanding of the organization's OT and IT architecture. Of particular importance is how the factory automation layer (OT layer)

Table 5.4 Typical digitalization task force team in a food & beverage manufacturing company

Personnel position	Typical tasks	Involvement timeline
Chief Digital Officer (CDO) / Chief Information Officer (CIO)	Defines digital transformation strategy, aligns with business objectives	Initial Phase, Ongoing
Digital Transformation Lead	Oversees the digitalization process, collaborates with IT and OT teams	Initial Phase, Ongoing
Manufacturing Process Engineer	Ensures automation and digitization align with production line efficiency	Ongoing, Full-Scale Implementation
Food Safety & Quality Expert	Implements digital tracking of quality control and food safety standards	Ongoing
Data Scientist / AI Specialist	Develops AI models for demand forecasting, predictive maintenance, and quality optimization	Ongoing, Full-Scale Implementation
SCADA & MES Specialist	Integrates Manufacturing Execution Systems (MES) with SCADA and ERP platforms	Initial Phase, Ongoing
Cybersecurity Expert	Implements secure digital infrastructure to protect production data and intellectual property	Initial Phase, Ongoing
IoT & Automation Engineer	Deploys IoT sensors for process monitoring, optimizes production control	Ongoing, Full-Scale Implementation
Supply Chain & Logistics Manager	Digitalizes supply chain tracking, optimizes warehouse & inventory management using AI	Full-Scale Implementation
Sustainability & Energy Manager	Uses digital tools to optimize energy consumption, waste management, and environmental compliance	Ongoing, Full-Scale Implementation
Maintenance & Reliability Engineer	Implements predictive maintenance, reduces downtime using machine learning analytics	Ongoing, Full-Scale Implementation

is connected to the business automation and information network (IT layer). A distributed control system (DCS) may consist of plants that are entirely isolated (air-gapped) from any form of external IT communication. In some cases, only externally directed communications with business IT layers are permitted through firewalls, while other plants allow limited internally directed communications through appropriate service connections.

In the ongoing conflict between IT and OT, intensified by digitalization in many industry sectors, many IT proponents consider the OT layer to be built on "legacy" hardware and software architectures. Such terminology may suggest that OT hardware requires replacement, but this assumption is not entirely accurate. A successful digital transformation in the process industry relies on achieving synergy between OT and IT layer software communication protocols, ensuring a seamless and secure data flow between the factory floor and the corporate network.

Integrating OT and IT layers is a complex task. Terms such as demilitarized zone (DMZ) and firewall are commonly used to describe the separation of the control network (OT

5.5 OT/IT Integration Assessment

layer) from the business network. One of the key objectives of digital transformation is to establish seamless and secure information transfer pathways between these layers. Achieving this often requires overcoming deeply ingrained silos and traditional mindsets. Distributed control systems with internet-based connectivity are already well established in several critical industrial sectors, such as the electricity grid.

Concerns regarding data and cybersecurity can present significant obstacles when connecting the OT and IT layers. The risks of network intrusion, hacking, data breaches, and other security threats are unavoidable realities in the IT domain. However, these risks can be mitigated through well-implemented security measures. Advanced tools for cybersecurity are increasingly available, enabling organizations to manage risk effectively. Completely isolating the factory floor due to cybersecurity concerns is comparable to avoiding travel out of fear of accidents—it may offer a sense of security but significantly limits potential benefits and opportunities.

5.5.1 Integration Considerations

Operational Technology (OT) in process industries involves real-time control systems, including sensors, actuators, PLCs, and RTUs, while Information Technology (IT) manages enterprise-level data storage, processing, and cloud computing. Integrating these layers enables seamless bidirectional data flow, enhancing efficiency and decision-making capabilities.

Hierarchical Network Architecture

The integration follows a multi-tiered architecture:

- **Field Level**: Sensors, actuators, and field instruments connected to PLCs and RTUs.
- **Control Level**: SCADA systems managing data acquisition and process automation.
- **Supervisory Level**: Manufacturing Execution Systems (MES) overseeing operations.
- **Enterprise Level**: IT infrastructure managing business intelligence and cloud integration.
- **Cloud/Fog/Edge Computing**: Advanced analytics, AI, and big data processing.

5.5.2 Hardware and Software Considerations

Hardware Selection Criteria

- **Field Devices**: Sensors with digital/analog outputs, actuators, smart transmitters.
- **Communication Gateways**: PLCs, RTUs, and edge computing devices.

- **Servers and Storage**: On-premises data centers and cloud servers.

Software, Protocols, and Firmware

Common protocols for OT-IT integration include:

- **Field Communication**: Modbus, HART, Profibus, OPC-UA.
- **Industrial Networks**: Ethernet/IP, PROFINET, MQTT for cloud integration.
- **Data Management**: SCADA historians, SQL/NoSQL databases, and cloud platforms.

Bandwidth Estimation

A key factor in integration is the required transmission bandwidth, calculated using:

$$B = N \times R \times (1 + O) \tag{5.1}$$

where:

- B = required bandwidth (bps),
- N = number of connected devices,
- R = data rate per device (bps),
- O = protocol overhead factor.

For example, if a plant has 500 sensors transmitting at 10 kbps each with a 20% overhead, the bandwidth required is:

$$B = 500 \times 10,000 \times 1.2 = 6 \text{ Mbps} \tag{5.2}$$

Cybersecurity Considerations

Ensuring secure communication and data access requires:

- Authentication and access control using role-based permissions.
- Network segmentation and firewalls to protect critical infrastructure.
- Data encryption for secure transmission and storage.
- Regular security audits and firmware updates to mitigate vulnerabilities.

The integration of OT and IT in process industries enhances efficiency, real-time monitoring, and decision-making capabilities. A well-structured network with appropriate hardware, communication protocols, and robust cybersecurity measures ensures a reliable and scalable digital transformation.

Table 5.5 Comparison of conventional DCS sensor topology versus IIoT sensor topology

Feature	Conventional DCS sensor topology	IIoT sensor topology
Sensor type	Wired analog/digital sensors with dedicated signal transmission	Wireless smart sensors with built-in processing and network capabilities
Communication Protocols	Proprietary industrial fieldbus protocols (e.g., HART, Modbus, PROFIBUS)	Standardized IoT protocols (e.g., MQTT, CoAP, OPC UA)
Data Transmission	Centralized, real-time deterministic control	Distributed, event-driven, and cloud-enabled
Processing Location	DCS controllers process data locally	Edge computing enables local processing, reducing latency
Scalability	Limited scalability due to physical wiring constraints	High scalability with wireless mesh networks and edge/cloud integration
Maintenance	Requires periodic manual calibration and maintenance	Remote monitoring and predictive maintenance using AI/ML
Cybersecurity	Closed-loop systems with limited external connectivity	Open connectivity requires advanced cybersecurity measures such as encryption and multi-factor authentication

5.5.3 Integration of IoT Devices in the OT Layer

The advent of the Industrial Internet of Things (IIoT) has introduced new possibilities for enhancing conventional process plant operational technology (OT) layers. Traditionally, Distributed Control Systems (DCS) have been the backbone of process industries, relying on wired sensor networks, centralized controllers, and predefined communication protocols. IIoT, in contrast, leverages a distributed, interconnected network of smart sensors, edge computing, and cloud-based analytics to provide greater flexibility, scalability, and advanced data-driven decision-making. Table 5.5 provides a comparison and complementarities between conventional DCS and IIoT based plant monitoring approaches.

5.6 Internet of Things

The term Internet of Things (IoT), coined originally at the turn of the century (1999), refers to a network of interconnected physical devices that communicate and exchange data through the internet. These "things" can range from everyday household items like smart thermostats, refrigerators, and wearable devices to industrial equipment such as sensors, machines, and even vehicles. Each IoT device is embedded with sensors, software, and communication technologies that allow it to collect and exchange data.

The key components of an IoT device are the actual devices (sensors or actuators) which are physical objects that monitor and control various conditions (*e.g.*, temperature, movement, humidity) of the physical realm. These devices are embedded with a layer of hardware and software that imparts the property of "connectivity" to these things. IoT devices use various communication methods, such as, Wi-Fi, Bluetooth, cellular networks, and LoRaWAN to send collected data to central systems. The IoT devices provide rudimentary platforms for analysis and processing of the collected data, including self-diagnostic capabilities. The final component of an IoT system is the user interface, which allows direct communication between the IoT device and other remote devices such as mobile apps, dashboards, or automated systems.

IoT devices can automate processes by sensing their environment and making decisions without human intervention. These devices can work together to create an integrated network, enabling smarter decision-making. finally, IoT generates large amounts of data, which can be analyzed to optimize performance, predict failures, and drive innovation.

Presently, IoT based systems are ubiquitous in smart homes and security systems, healthcare for monitoring patient health or remote monitoring of health conditions of individuals. At a consumer level IoT based devices combined with mobile phones and applications have pervaded our daily lives, and are ubiquitous. These devices are also revolutionizing the industrial sector, with industrial IoT (IIoT) systems replacing conventional localized process control systems. Smart sensors with IoT generated data analysis are increasingly being used for predictive maintenance in the manufacturing environments. Ability to distribute a large number of sensors with low cost communication has allowed these systems to gain traction in smart agriculture (monitoring soil conditions, water usage and crop health monitoring), as well as in smart cities allowing traffic flow management, waste management, and energy efficiency gains (smart lighting and district heating or cooling).

IoT devices have changed the cost structure of industrial automation by making the data acquisition and transmission considerably inexpensive. A large array of sensory and motor devices can be connected to a network, and can communicate through TCP/IP, UDP/IP, CANbus or other communication protocols. These protocols can achieve communication between devices through wired (LAN), wireless, or cellular communication. IoT devices can communicate directly with other IoT devices, as well as with a central hub. Therefore, a large and relatively robust sensory network can be established to monitor different sections of a manufacturing facility. The software communication framework of these IoT devices are also fairly standardized and significantly less expensive as these can use conventional TCP/IP protocols for network based communications. Together, IoT devices as well as new generation ethernet based communication protocols have made acquisition of data and communication between devices over the internet quite straightforward and mainstream.

5.6 Internet of Things

The ease of communication between devices through the internet also contributes to one of the major risks of IoT devices, namely, their vulnerability to cyberattacks, as well as privacy concerns over the data they generate. As the industrial automation sector is still transitioning from the dated process control hardware, direct wired data transfer protocols and fieldbus technologies (such as HART, Profibus, Modbus, etc.), there exist considerable heterogeneity in low level hardware, firmware, and software. Devices from different manufacturers may not always communicate seamlessly. Although (Ethernet/IP) EIP protocol is shaping up as a commonly available protocol offered by many brands of operating technology (OT) hardware, ethernet based protocols are still not considered sufficiently secure, or deterministic, as opposed to the traditional fieldbus technologies (such as Profibus) for many time critical (real-time) control scenarios.

Integrating IIoT-based sensors alongside conventional DCS architectures presents both challenges and opportunities. While DCS architectures are designed for deterministic control with real-time constraints, IIoT architectures provide enhanced data analytics, predictive maintenance, and remote monitoring capabilities. The integration of these two approaches allows for improved operational efficiency, but it requires a well-structured communication and networking framework to ensure seamless interoperability.

Key Value Propositions of Augmenting Conventional Sensor Networks with IIoT-Based Sensing

Integrating IIoT sensors alongside conventional process instrumentation introduces several advantages, including:

- **Enhanced Data Availability**: IIoT sensors provide continuous, high-frequency data streams, enabling real-time insights and analytics.
- **Predictive Maintenance**: Smart sensors with built-in analytics detect early signs of equipment degradation, reducing unplanned downtime.
- **Remote Monitoring**: Wireless connectivity allows engineers and operators to monitor critical parameters from remote locations, improving decision-making and reducing site visits.
- **Flexibility and Scalability**: IIoT-enabled sensors can be deployed without extensive wiring, making it easier to expand monitoring capabilities in large-scale plants.
- **Integration with Cloud and Edge Computing**: IIoT devices can leverage cloud-based AI and machine learning algorithms to optimize process efficiency and fault detection.
- **Energy Efficiency and Cost Reduction**: Battery-operated wireless sensors minimize infrastructure costs associated with wiring, power distribution, and hardware maintenance.

The convergence of traditional DCS architectures with IIoT-based sensing creates a hybrid model that balances reliability, real-time control, and data-driven analytics. This integrated approach maximizes process efficiency, enhances operational intelligence, and enables industries to transition towards Industrie 4.0 with minimal disruption.

5.6.1 Communication Protocols

In an industrial digital transformation framework, seamless network communication between various databases spanning from the Remote Terminal Unit (RTU) level to the cloud requires a combination of robust networking tools and standardized communication practices. At the RTU level, field devices and controllers typically communicate using industrial protocols such as Modbus, PROFIBUS, and HART over wired or wireless connections. These data streams are aggregated at the Programmable Logic Controller (PLC) or Distributed Control System (DCS) level, where OPC UA (Open Platform Communications Unified Architecture) serves as a key middleware standard, enabling secure, interoperable data exchange between industrial automation devices and higher-level IT systems. Moving upwards, edge computing nodes and industrial gateways play a crucial role in preprocessing, filtering, and transmitting data to on-premise or cloud-based databases, using lightweight protocols such as MQTT (Message Queuing Telemetry Transport) or AMQP (Advanced Message Queuing Protocol). For large-scale data storage and real-time analytics, cloud-based solutions like AWS IoT Core, Microsoft Azure IoT Hub, or Google Cloud IoT provide scalable environments for hosting data lakes and AI-driven analytics. Cybersecurity practices such as VPNs, firewalls, data encryption (TLS/SSL), and role-based access control (RBAC) ensure data integrity and protection across the network layers. Efficient bandwidth management, Quality of Service (QoS) policies, and redundancy mechanisms (such as dual Ethernet and 5G failover) further enhance reliability, ensuring uninterrupted communication from plant-floor databases to enterprise-level cloud infrastructures.

In a distributed edge, cloud, and fog computing system, database communication and connection protocols must ensure seamless data exchange, security, and real-time access across different computational layers. The key protocols used in such systems include:

Message-Oriented Middleware (MOM) Protocols

These lightweight, efficient protocols are widely used for IoT and distributed computing:

- MQTT (Message Queuing Telemetry Transport): A publish-subscribe protocol optimized for low-bandwidth, high-latency networks, commonly used in IoT edge-to-cloud communication.
- AMQP (Advanced Message Queuing Protocol): Ensures reliable, asynchronous communication with message brokers, suitable for cloud-based applications.
- DDS (Data Distribution Service): A high-performance, real-time data exchange protocol used in industrial automation and robotics.

Industrial Protocols for Edge and Fog Connectivity

These protocols enable industrial devices and systems to communicate:

- OPC UA (Open Platform Communications Unified Architecture): A platform-independent protocol ensuring secure and standardized data exchange between industrial automation systems and cloud platforms.
- Modbus TCP/IP & PROFIBUS/PROFINET: Used for field device communication, particularly in SCADA and PLC-based systems.
- BACnet (Building Automation Control Network): Commonly used in facility automation and industrial HVAC systems.

Database Connection Protocols

These protocols facilitate database interactions across edge, fog, and cloud layers:

- ODBC (Open Database Connectivity): A standard API for accessing SQL databases such as MySQL, PostgreSQL, and Microsoft SQL Server.
- JDBC (Java Database Connectivity): A Java-based API used for database access in enterprise applications.
- RESTful APIs (Representational State Transfer): Used for cloud-based database interactions, allowing applications to send HTTP requests to access or update data.
- RPC (Remote Procedure Call): A high-performance protocol useful for micro-services and database interactions in distributed systems.

Cloud and Web-Based Communication Protocols

These protocols support secure and efficient cloud database connectivity:

- HTTPS (Hypertext Transfer Protocol Secure): Encrypts database queries over the web using TLS.
- WebSockets: Enables persistent, real-time, bidirectional communication between edge devices and cloud applications.
- CoAP (Constrained Application Protocol): A lightweight alternative to HTTP, optimized for IoT and constrained networks.

Edge-Fog-Cloud Data Synchronization Protocols

These are used to maintain consistency across distributed databases:

- Event Streaming Platforms: Distributed (such as kafka) and cloud-native (Apache Pulsar) event-streaming platforms for real-time data ingestion between edge, fog, and cloud layers, and optimization of messaging and event streaming for large-scale deployments.
- Time-Series Databases (TSDBs): Used to store time-sensitive sensor data from edge and fog nodes.

The communication between different databases, whether requiring automated server to server API calls, or user initiated queries, must be secure, with authentication tools in place. These tools ensure secure database connections in distributed architectures:

- TLS/SSL (Transport Layer Security/Secure Sockets Layer): Encrypts data transmission between databases and applications.
- OAuth 2.0 and OpenID Connect: Secure authentication and authorization for accessing cloud databases.
- Zero Trust Architectures: Implement access control and security policies for edge-fog-cloud networks.

By leveraging these protocols, a distributed edge-fog-cloud system ensures efficient, secure, and scalable database communication, enabling real-time data processing and decision-making across industrial and enterprise applications.

5.6.2 Cybersecurity and Authentication Protocols

Digital transformation introduces new technological advancements but also expands the attack surface for cyber threats. Cybersecurity tools play a crucial role in securing enterprise networks, data, and operational technology (OT) during digital transformation. This section provides a brief summary of cybersecurity tools and their implementation in an enterprise setting.
Categories of Cybersecurity Tools To ensure comprehensive protection, enterprises deploy multiple layers of cybersecurity tools. These can be categorized as follows:

- **Network Security Tools**: Firewalls, Intrusion Detection Systems (IDS), and Intrusion Prevention Systems (IPS) help monitor and control network traffic.
- **Endpoint Security Tools**: Antivirus software, endpoint detection and response (EDR) systems, and mobile device management (MDM) solutions secure user devices.
- **Identity and Access Management (IAM)**: Multi-factor authentication (MFA), single sign-on (SSO), and role-based access control (RBAC) ensure secure user authentication and access control.

- **Data Protection Tools**: Encryption mechanisms, data loss prevention (DLP) tools, and secure backup solutions safeguard sensitive enterprise data.
- **Threat Intelligence and Monitoring**: Security Information and Event Management (SIEM) systems aggregate logs and detect threats in real time.
- **Cloud Security Tools**: Cloud Access Security Brokers (CASB), cloud workload protection platforms (CWPP), and secure web gateways protect cloud environments.

The effective implementation of cybersecurity tools requires a structured approach:

1. **Risk Assessment and Security Frameworks**
   Enterprises should conduct a comprehensive risk assessment to identify vulnerabilities in their IT and OT infrastructure. Security frameworks such as NIST Cybersecurity Framework, ISO/IEC 27001 [4], and IEC 62443 [5] for industrial control systems provide structured guidelines for implementing cybersecurity.
2. **Zero Trust Security Model**
   A Zero Trust approach ensures that no entity is trusted by default, even within the network perimeter. This involves continuous verification of user identities, strict access controls, and micro-segmentation of networks.
3. **Secure Network Architecture**
   Network segmentation should be enforced to separate IT, OT, and cloud environments. Implementation of software-defined perimeters (SDP) helps protect against lateral movement of cyber threats.
4. **Incident Response and Security Operations Center (SOC)**
   Enterprises must establish an incident response plan and deploy a SOC to monitor security threats 24/7. Automated response systems and forensics tools aid in quick mitigation of security incidents.
5. **Continuous Monitoring and Compliance**
   Regular penetration testing, security audits, and compliance tracking ensure adherence to industry regulations. AI-driven threat detection and anomaly detection systems improve security postures.

Cybersecurity Challenges in Digital Transformation Despite implementing advanced security measures, enterprises face several challenges:

- **Integration Complexity**: Legacy systems may not seamlessly integrate with modern security tools, requiring additional customization.
- **Scalability**: As enterprises grow, cybersecurity solutions must scale to accommodate increased data traffic and endpoints.
- **User Awareness and Training**: Human error remains a significant vulnerability. Regular security training programs help mitigate risks associated with phishing and social engineering attacks.

- **Compliance Management**: Different industries must comply with distinct cybersecurity regulations (e.g., GDPR, HIPAA, CMMC), requiring tailored security measures.

Cybersecurity is a critical enabler of digital transformation. By deploying a layered security approach, leveraging AI-driven security analytics, and enforcing a Zero Trust model, enterprises can effectively mitigate cyber risks. Ongoing monitoring, employee training, and regulatory compliance play a vital role in ensuring a secure digital enterprise. Future trends, such as quantum-safe cryptography and AI-powered autonomous security, will continue to shape the cybersecurity landscape.

5.7 Data Structures, Algorithms, and Automation

Once all information pathways have been mapped out and connections established, the next step involves setting goals and assessing the requirements for data management, and processing.

A careful evaluation of current data usage and information flow within the organization is essential to determine whether digitalization will provide added benefits. Excessive data accumulation can often lead to information paralysis. Therefore, it is necessary to approach the integration of big data, analytics, predictive analytics, regression, classification, machine learning, and artificial intelligence with caution.

An undue emphasis on data acquisition can place an additional burden on the organization, as well as on its technically proficient staff and management, in terms of processing and interpreting the data. Many organizations have undergone digital transformation through the implementation of enterprise resource planning (ERP) systems, typically aimed at automating business processes through data management. When properly executed, such implementations are highly valuable and transformative. However, imperfect and time-consuming ERP system deployments are common, often resulting in significant challenges during the transition period for businesses.

Converting data to knowledge and wisdom is achieved through suitable computational algorithms. An important aspect of digitalization is the association of appropriate algorithms to process and analyze the accumulated data. The selection of algorithms for digital modeling and analysis of physical plants depends on multiple factors, including the nature of the data, computational constraints, and the desired balance between data-driven and physics-driven approaches. Key features of data that influence algorithm selection include its volume, variety, veracity, and velocity. High-dimensional datasets, common in industrial settings, require dimensionality reduction techniques, which can be statistical, such as Principal Component Analysis (PCA), or physics based, such as similarity or dimensional analysis, to extract relevant information while reducing computational complexity. Overfitting, a major challenge in data-driven modeling, occurs when an algorithm learns noise instead of true patterns, leading to poor generalization. Cost factors for developing and maintaining complex com-

5.7 Data Structures, Algorithms, and Automation

putational models—especially deep learning—include the need for large labeled datasets, high-performance computing resources, and specialized expertise. The trade-off between interpretability and predictive power must also be carefully managed, as deep learning models often function as black boxes, whereas physics-based models provide explainability rooted in established engineering principles.

When planning a digital transformation, these criteria must be evaluated to ensure a sustainable and efficient implementation. For example, in predictive maintenance or process optimization, a hybrid approach combining physics-based models with machine learning can enhance accuracy while maintaining interpretability. The computational costs of training deep learning models must be weighed against their incremental performance gains, and simpler models should be prioritized when real-time decision-making is required. Additionally, data governance strategies must be established to manage data quality, security, and compliance, ensuring that digital models remain reliable over time. Finally, scalability and adaptability should be considered, as industrial systems evolve with new sensors, operational changes, and business goals. A well-planned algorithm selection strategy will determine whether the digital transformation initiative enhances decision-making and operational efficiency or results in unsustainable complexity and cost overruns.

The implementation of business automation should be assessed by evaluating the staff's acceptance of new workflows and tools. It is important to determine whether employees are overwhelmed with extensive data processing tasks, leading to an excessive reliance on spreadsheets. Additionally, the extent to which calendars, meetings, presentations, and dashboards operate autonomously should be examined to ensure that employees are not merely following instructions and notifications generated by automated systems.

In a well-digitalized environment, it should not be necessary to manually download data from a database into spreadsheets, perform calculations, create charts, and use these calculations to prepare presentations, dashboards, and reports. Instead, calculations should be performed automatically, reports should be generated seamlessly, and dashboards should be readily accessible to stakeholders, allowing for data sharing and real-time information analysis. This implies that a digitalization system should function as an embedded computational framework that automates these tasks, thereby reducing the repetitive workload of human stakeholders.

5.7.1 Computational Philosophy and Infrastructure for Digital Transformation

The digital transformation of industrial enterprises necessitates a robust computational framework that serves as the nervous system of the plant, ensuring seamless bidirectional communication between operational technology (OT) layers and cloud-based computational resources. This section explores the fundamental principles governing the computational philosophy, hardware, software, networking strategies, and programming approaches necessary to implement an effective digital transformation strategy.

Computational Philosophy and Design Principles

A well-designed computational infrastructure for digital transformation must be structured to facilitate three primary objectives:

1. **Real-time Process Control**: Edge-level computing must be capable of executing low-latency control loops that dynamically respond to plant operations, akin to synaptic joints in biological nervous systems.
2. **Higher-Order Decision Making**: Cloud and fog computing resources should synthesize data into actionable insights, leading to optimization, predictive maintenance, and anomaly detection.
3. **User-Centric Interactions**: The system must allow multiple levels of user engagement, from plant operators monitoring real-time dashboards to data scientists querying historical databases for analysis.

To achieve these goals, digital transformation systems must integrate multiple computational layers, leveraging distributed computing paradigms to maintain both low-latency and high-throughput processing capabilities.

Hardware Considerations

The hardware selection for digital transformation systems should be based on a hierarchical structure:

- **RTU/PLC Level**: These must be equipped with real-time operating systems (RTOS) to handle deterministic control.
- **Edge Computing Nodes**: Industrial PCs, microcontrollers, or embedded systems must be deployed to preprocess data, execute control logic, and interface with cloud platforms.
- **Fog Computing Infrastructure**: Intermediate computational resources (such as on-premise data centers) facilitate local analytics, reducing dependency on cloud-based processing.
- **Cloud Computing Resources**: High-performance servers and distributed data lakes provide scalable storage, machine learning model execution, and global process optimization.

Software and Programming Principles

The software architecture must be designed to allow interoperability across different computational layers. Key programming paradigms include:

5.7 Data Structures, Algorithms, and Automation

- **Event-Driven Programming**: Enables asynchronous responses to real-time data changes.
- **Microservices Architecture**: Allows modular deployment of services that handle different process functionalities.
- **Secure API Interfaces**: Ensure seamless data exchange while enforcing cybersecurity policies.
- **Edge-AI Integration**: Enables inferencing and decision-making at the edge without continuous cloud reliance.

Software development should also leverage containerization tools to facilitate portability and maintainability across different environments.

Data Communication and Networking Strategies

A robust networking strategy is critical to ensuring efficient data transmission between plant-level sensors and cloud-based analytics. Key networking elements include:

- **Time-Sensitive Networking (TSN)**: Provides deterministic latency for real-time control applications.
- **MQTT and OPC UA Protocols**: Enable secure and lightweight data exchange across industrial components.
- **5G and LPWAN Integration**: Support high-bandwidth and long-range data transmission for remote industrial assets.
- **Edge and Cloud Data Synchronization**: Ensures consistency and reliability in historical and real-time data analytics.

Bandwidth estimation should be based on the volume of sensor data, polling frequency, and redundancy mechanisms required to maintain system resilience. The Nyquist criterion should be employed to determine optimal sampling frequencies for different process parameters to avoid data loss or aliasing.

Cybersecurity and Human-System Interaction

A fully digitized industrial system must incorporate robust cybersecurity measures. As discussed earlier, role-based access control (RBAC), encryption, and secure protocols are required elements to restrict unauthorized user access, ensuring data integrity, and confidentiality during transmission. Furthermore, **Anomaly Detection Algorithms** are used to identify potential security breaches or process failures.

Human interactions with the cyber-physical system should be facilitated through:

- **HMI Dashboards**: Provide real-time visibility into process performance.

- **Natural Language Processing (NLP) Interfaces**: Allow operators to query systems intuitively.
- **Augmented Reality (AR) Integration**: Enhances on-site maintenance by overlaying digital process information.

The digital transformation of industrial enterprises requires a computational framework that integrates real-time control, distributed analytics, and user interaction. By employing edge computing for rapid feedback loops and cloud computing for high-level decision-making, enterprises can build a robust digital nervous system that ensures operational efficiency, reliability, and security. The successful implementation of this architecture depends on selecting the right hardware, designing interoperable software, and establishing secure communication networks that can seamlessly interconnect plant-floor devices with enterprise-wide digital infrastructure.

5.8 Human-Machine Interfaces in a Digitally Transformed Enterprise

The integration of digital technologies within industrial enterprises has led to the emergence of cyber-physical systems that transform traditional plant operations. In a digitally transformed enterprise, human-machine interfaces (HMIs) serve as the critical bridge between operators, engineers, and business stakeholders, enabling seamless interaction with the plant. Unlike conventional systems that rely on manual intervention, modern HMIs facilitate an environment where data-driven insights, predictive analytics, and autonomous decision-making redefine operational paradigms.

A digitalized smart-plant may be designed to provide the intended users (stakeholders) abilities:

1. To monitor all operational aspects of the plant in real-time. This may include statistics about how long the plant has been operating steadily, what is the production rate, and other relevant vital information about the plant.
2. To receive information from the plant regarding any impending changes in influent (raw material) quality or quantity, operating conditions, as well as problems with the component processes.
3. Process information about optimal conditions for production and how the plant operating conditions can be altered to meet such optimized production goals.

When designing user interfaces for a digitalization project, the primary focus should be on delivering information that adds value beyond what is already available through existing sources. The interaction between the user and the digitalized plant should be more intelligent and meaningful, enabling deeper insights (knowledge) and predictive capabilities (wisdom) regarding the manufacturing process. An effective user interface (UI) should empower users to quickly address higher-level analytical questions such as "how" and "why" events occur, rather than merely presenting raw data on "what happened" within the plant.

5.8.1 The Cyber-Physical System Paradigm

A fully digitalized plant functions as a cyber-physical system (CPS), where physical assets, sensors, controllers, and software-defined intelligence create a dynamic operational environment. This ecosystem enables real-time monitoring, automated diagnostics, and self-optimizing processes, reducing human workload while improving efficiency and sustainability. Instead of responding to alarm conditions or scheduled maintenance tasks, operators engage with systems that proactively suggest interventions based on predictive analytics.

A significant portion of digitalization efforts is often directed toward creating user interfaces (UI), particularly dashboards, which provide little to no substantial value. In most cases, plant SCADA systems already include dashboards with alarms and alerts configured within them. Therefore, developing additional dashboards is generally inefficient and unnecessary unless they offer functionalities beyond what is already available in SCADA control rooms. Many companies invest considerable resources in designing user interfaces and providing operators with tablets or other gadgets. However, carrying such devices on the plant floor often poses safety hazards for both the operator and the equipment. Additionally, these dashboards are rarely utilized once the initial novelty fades. If dashboards are developed, tracking their usage can be a valuable practice to assess user engagement and interaction patterns.

5.8.2 Shifts in Human-Machine Interaction

As industrial automation advances, the role of human operators transitions from reactive intervention to proactive decision-making. Future smart plants will witness the following changes in human-machine interactions:

- **From Alarm-Driven to Proactive Maintenance**: Traditional maintenance relied on periodic schedules or operator-triggered interventions. In a smart plant, maintenance is executed on an as-needed basis, with the system generating real-time requests based on degradation models, condition monitoring, and AI-driven diagnostics.
- **Work Schedules and Workflows**: Operators will no longer perform routine data collection and manual inspections. Instead, they will oversee decision support systems that guide maintenance actions, asset utilization, and production adjustments. Workflows will evolve from manual logging and shift-based reporting to continuous digital oversight and automated workflow management.

- **Enhanced Interaction Tools**: Human-machine interaction will increasingly rely on agent-based communication tools, augmented reality (AR) overlays for visual diagnostics, and virtual reality (VR) training modules for skill development. Instead of interpreting raw sensor data, personnel will interact with contextualized insights presented through intuitive dashboards.

5.8.3 Decision Support and Operational Setpoints

In a digitally transformed enterprise, operational setpoints and control strategies will no longer be solely determined by static process models but will be dynamically adjusted by decision support systems (DSS). These systems leverage real-time data, historical trends, and optimization algorithms to recommend setpoint changes that enhance efficiency, reduce energy consumption, and maximize yield. Operator roles will evolve to focus on validating and fine-tuning these recommendations rather than manually setting process parameters.

5.8.4 Business Stakeholder Interactions with Digital Tools

Business stakeholders, including executives and analysts, will benefit from agent-based digital assistants that streamline access to operational data. Unlike traditional enterprise systems requiring manual data queries and custom report generation, modern digital frameworks will feature conversational AI agents capable of understanding stakeholder requirements and autonomously retrieving and analyzing relevant data. This shift will:

- Eliminate the need for manual data extraction, formatting, and visualization.
- Reduce the time required for decision-making by presenting real-time insights.
- Enable scenario analysis through automated simulation and forecasting models.
- Foster a data-driven corporate culture where strategic decisions align with predictive and prescriptive analytics.

5.8.5 Eliminating Manual Historical Analysis and Enabling Continuous Computing

One of the most profound transformations in the digitally enabled enterprise is the transition from retrospective data analysis to real-time decision-making. Traditional industrial workflows involved downloading process data into spreadsheets for offline analysis and goal-seeking. In contrast, an always-on computing infrastructure continuously aggregates, synthesizes, and contextualizes data, ensuring that insights are readily available without human intervention. This architecture enables:

- Automated anomaly detection, reducing reliance on historical trend analysis.
- Continuous performance optimization through adaptive control algorithms.
- Automated data, information, and knowledge (DIK) synthesis, allowing human personnel to focus on higher-order strategic improvements rather than raw data interpretation.

A more critical aspect of digital transformation is delivering an intuitive, well-designed, and practical user experience (UX) that enhances the overall digitalization initiative. The objective is to delegate data collation, assimilation, and analysis tasks to computerized systems embedded within the digitalized plant, thereby reducing the manual workload for operators. This delegation is the core principle of automation. Ideally, the user experience should be centered around responding to automated notifications and alerts generated by the digital support system and scheduling physical actions—such as maintenance tasks—with greater confidence and efficiency. Furthermore, digital transformation should not impose additional burdens on human operators but should instead empower them with enhanced control over plant operations, accessible from anywhere and at any time.

The human-machine interface in a digitally transformed enterprise redefines the interaction between personnel and industrial systems. With advanced decision support, proactive maintenance, real-time computational intelligence, and agent-based analytics, the digitalized plant shifts human roles from routine operational oversight to strategic decision-making. This transformation not only enhances efficiency and sustainability but also fundamentally improves the quality of life for plant operators and business stakeholders, making industrial operations more intelligent, responsive, and resilient.

5.9 Assessment Questions

A list of typical questions for consideration during the initial conception and planning of a digital transformation project is provided below.

> *1. What is the digitalization end-game?*
> *The objectives and key results that quantitatively indicate the achievement of these objectives should be clearly defined. The focus should remain on the specific manufacturing process, working conditions, and operational environments of the plant while developing these objectives. An assessment of the current status of the manufacturing environment is essential to determine the desired level of advancement. Every plant and organization operate with distinct workflows, making it crucial to consider these specifics.*

2. What types of data should be collected, and how frequently?
Data collection involves costs related to storage, processing, and retrieval. As database size increases, analysis and processing become more complex, expensive, and time-consuming. Additionally, indirect costs must be considered, whether data is stored on the cloud or on local hardware, as data processing and associated computations consume energy and generate heat. Increased energy consumption and cooling requirements contribute to greenhouse gas (GHG) emissions, potentially conflicting with a business's sustainability goals.

3. How should the data be processed?
The emphasis should be on enhancing knowledge creation and wisdom generation from data rather than replicating existing functionalities. Automation of data processing activities should be prioritized, with less focus on dashboard creation.

4. What questions should be addressed using the data?
The identification of key stakeholders and the potential benefits they may derive from the insights gained through data analysis is essential. Mapping out workflows during the planning phase and implementing measures to streamline these processes will improve operational efficiency. Additionally, provisions for flexibility should be considered, as plant conditions will evolve over time, and workforce transitions may occur. Business continuity planning principles should be incorporated into the digitalization architecture.

5. How should responses to these questions be obtained?
Consideration should be given to querying databases, aggregating and representing data to display results, and defining actions based on data-driven insights. This includes assessment of pertinent analysis algorithms, as well as approaches of representing the model of the physical plant into the digital realm. Furthermore, the communication of this information to appropriate stakeholders must be strategically planned.

6. Should data probing be conducted manually, at automated intervals, or triggered under specific conditions?
Ideally, new methods of querying data should be established to synthesize knowledge and wisdom as the plant and operating environment evolve. Once a manual framework for this process has been developed, the digital support system should be capable of autonomously executing these tasks on a recurring basis.

7. What is the expected value of the knowledge or wisdom extracted from the data, and how can this value be quantified?
The key metrics for measuring success include stakeholder adoption and tangible benefits in terms of plant economics, operational efficiency (e.g., reductions in OpEx,

energy, and resource consumption), reliability improvements, and the achievement of corporate sustainability goals. Incorporating these evaluation systems into the digital transformation framework will facilitate the development of a truly smart, agile, and adaptive manufacturing environment.

References

1. Rob J. Hyndman and George Athanasopoulos. *Forecasting: Principles and Practice*. OTexts, Melbourne, Australia., 3 edition, 2021. https://otexts.com/fpp2/
2. Gang Kou, Yi Peng, and Gang Wang. Application of noise-filtering techniques to data-driven analysis of industrial processes. *Journal of Process Control*, 105:45–58, 2023. https://www.sciencedirect.com/science/article/pii/S0142061523007780
3. Wei Zhang and Ming Li. Tsn time synchronization based on kalman filtering. *Journal of Network and Computer Applications*, 8(6):55–61, 2023.
4. International Electrotechnical Commission. ISO/IEC 27001: Information security management systems — Requirements. Technical report, International Organization for Standardization (ISO) & International Electrotechnical Commission (IEC), 2022.
5. International Electrotechnical Commission. IEC 62443: Industrial communication networks – IT security for networks and systems. Technical report, International Organization for Standardization (ISO) & International Electrotechnical Commission (IEC), 2023.

Implementing Digital Transformation

6.1 Digital Transformation: From Strategy to Execution

In the previous chapter, we outlined a comprehensive approach for enterprise-wide digital transformation planning. We discussed key focus areas, established the primary objectives of developing a digital transformation strategy, and identified the necessary resources, tools, and methodologies required for successful implementation. In this chapter, we shift our focus to the technical execution of a digital transformation strategy in the chemical process industry.

As discussed in Chap. 3, digital transformation can be integrated at any stage of a process plant's lifecycle. While implementing digital transformation in a newly constructed manufacturing plant may appear to be an attractive option, this approach carries significant risks. A digital-first, top-down design for a new manufacturing plant allows for seamless human-machine interactions in a fully digital realm. However, given that digital transformation is still an evolving paradigm, untested at scale in many industries, a cautious and balanced approach is often preferable.

For most enterprises, a pragmatic strategy is to first conduct a pilot implementation on an existing mid-life manufacturing unit. This allows the organization to test, refine, and validate the digital transformation framework before committing to a full-scale deployment across the plant or enterprise. A controlled pilot project provides invaluable insights into potential integration challenges, workforce adaptation requirements, and system interoperability issues while minimizing risks to ongoing production.

A well-chosen pilot unit, such as a representative production line, a process utility system (e.g., water purification), or a captive power generation facility, enables companies to gain first-hand experience with digitalization without jeopardizing full-scale operations. The lessons learned from such a brownfield pilot can then be systematically applied to future greenfield projects. This ensures that the next generation of plant design is optimized for automation, advanced analytics, and adaptive intelligence, laying a robust foundation for a truly digital-first manufacturing environment.

6.2 Pilot Implementation in a Mid-Life Manufacturing Unit

To implement digitalization as a retrofit upgrade to an existing plant, one cannot readily envision a top-down approach, since it is expensive, time consuming, and disruptive to the operations of the plant. Such a capital infusion on older and depreciating plants may not be economically viable. Retrofit digitalization requires building on the existing automation framework of the plants. This bottom-up approach requires a thorough understanding of the processes and automation steps of the existing plant. This assessment of the existing control and automation framework of a plant is important to avoid conflicts between the existing framework and the new digitalization systems to be implemented. Such conflicts may manifest as redundancy, impaired reliability, and security. A thoughtful investment in this bottom up approach of implementing a digitalization retrofit can significantly improve capacity utilization, efficiency, reliability, and sustainability of existing process plants.

To implement a retrofit digitalization of a mid-life plant, an important consideration is to clearly articulate the goals to be achieved by the added digitalization features, assess the capabilities of the process plant's existing automation framework, perform a gap analysis, define key performance indicators (KPIs) and metrics to measure success, project expected returns on investment (ROI), and then embark on an improvement plan. Fundamentally, for such transformations, one has to first ask: *"What additional benefits does digitalization offer that is not already provided by the current plant automation?"*

6.2.1 Business Case for Retrofit Digital Transformation in Brownfield Plants

Digital transformation in brownfield process plants presents unique challenges and opportunities. Retrofitting a digital framework into an existing plant requires careful planning, goal setting, and a strong business case to justify the investment. This section provides a structured approach to defining objectives and making a compelling business case for digital transformation based on economic justification.

Economic Justification for Digital Transformation

Consider an existing process plant that is eight years old, operating with a legacy operational technology (OT) and information technology (IT) infrastructure. The plant lacks an integrated process data system, with no SCADA historian, and relies heavily on manual data collection, particularly in laboratory analyses and batch reporting. The limitations of this outdated system lead to inefficiencies, process inconsistencies, and an inability to leverage data-driven decision-making.

To evaluate the feasibility of digital transformation, we establish the direct and indirect costs and benefits associated with the transition. The cost elements include capital expenditure (CapEx) for upgrades to OT infrastructure, deployment of a SCADA historian, integration of laboratory information management systems (LIMS), and implementation of cloud or edge computing for analytics, operational expenditure (OpEx) for software licensing, cybersecurity enhancements, and workforce training, technology debt costs or the financial burden of operating outdated systems, requiring frequent maintenance and limiting future scalability, and transition and downtime costs.

Annualized Investment and Technology Debt in Digital Transformation

When undertaking a digital transformation project for a brownfield plant, it is essential to evaluate the financial impact in a structured manner. One of the key metrics in this evaluation is the **annualized investment**, which accounts for both capital and operational expenditures, retraining costs, and technology debt considerations. The total cost of digital transformation can be represented through an annualized investment model:

$$\text{Annualized Investment} = \text{CapEx} + \text{Annualized OpEx} \\ + \text{Retraining Costs} + (\text{Initial Tech Debt} \\ - \text{Tech Debt Reduction Over Time}) \quad (6.1)$$

where:

- **CapEx** (Capital Expenditure): The initial investment in new hardware, software, and infrastructure required for digital transformation.
- **Annualized OpEx** (Operational Expenditure): The recurring costs associated with maintaining and running the new digital infrastructure, including cloud services, cybersecurity, and software licensing.
- **Retraining Costs**: The cost of upskilling employees to operate within the new digital ecosystem, including training programs, workshops, and documentation.
- **Initial Tech Debt Cost**: The existing inefficiencies, maintenance costs, and legacy system constraints that must be managed during the transition.
- **Tech Debt Reduction Over Time**: The eventual savings and efficiency gains achieved as outdated systems are phased out and replaced by digital solutions.

Technology debt refers to the accumulated inefficiencies, outdated systems, and suboptimal technological decisions that create ongoing operational and maintenance costs. This debt arises due to:

- Dependence on legacy IT/OT systems that lack interoperability with modern digital platforms.
- Workarounds and patches needed to maintain outdated software and hardware.
- Security vulnerabilities in aging systems requiring costly protective measures.
- Inefficiencies in data collection and processing, leading to higher manual labor costs and slower decision-making.

While digital transformation ultimately *reduces* technology debt, in the short term, the transition incurs costs due to:

- Parallel operation of old and new systems during the transition phase.
- Workforce adaptation expenses as employees learn to operate new tools and workflows.
- The need for temporary integration solutions between legacy and digital systems.
- Accelerated depreciation or write-off costs for obsolete infrastructure.

A well-planned digital transformation initiative aims to significantly reduce technology debt by replacing outdated infrastructure with modern, scalable, and interoperable solutions. However, in the short term, addressing technology debt adds to the total cost of digitalization because existing inefficiencies and integration challenges must be resolved before achieving a streamlined digital ecosystem.

Although technology debt is an undesirable cost, its inclusion in the annualized investment calculation is essential to provide a realistic view of the total financial impact of digital transformation. By addressing technology debt early in the digitalization process, enterprises can prevent future escalations in maintenance and security costs, leading to long-term operational savings and improved agility.

Thus, while digitalization reduces technology debt over time, its immediate resolution requires investment. This highlights the importance of strategic financial planning in digital transformation, ensuring that costs are balanced against expected improvements in efficiency, productivity, and business resilience.

Benefits

The expected benefits from digital transformation include:

- **Production Cost Reduction:** Enhanced automation, predictive maintenance, and optimized process control will decrease energy consumption, reduce raw material waste, and minimize downtime.
- **Increased Production Volume:** Improved process monitoring and control will enable higher throughput better capacity utilization, with consistent product quality.
- **Operational Efficiency:** Reduction in human errors, streamlined workflow, and faster decision-making through real-time data insights.

- **Regulatory Compliance and Traceability:** Digital record-keeping enhances compliance with industry standards and reduces audit risks.

Key Performance Indicators (KPIs) and ROI Measurement

To track the success of digital transformation, the following KPIs may be established:

- Reduction in unplanned downtime (%)
- Improvement in overall equipment effectiveness (OEE)
- Reduction in process variability and defects
- Increase in production throughput (%)
- Reduction in manual data entry errors and reporting time (%)
- Cost savings from optimized energy and material usage ($)

Return on investment (ROI) can be calculated simply as:

$$\text{ROI} = \frac{\text{Annual Savings} + \text{Revenue Increase} - \text{Investment}}{\text{Investment}} \times 100\%. \tag{6.2}$$

where the Investment should be considered as annualized investment.

Irrespective of what economic models are used, what values are generated, or what ROIs are calculated, it should translate into some form of demonstrable, universally accepted before-and-after digital transformation metric for the plant through which the efficacy of the digital transformation will be assessed by the enterprise. Figure 6.1 schematically depicts a typical representation of such a performance metric (here, some annualized plant asset value) reflecting the ROI from Eq. (6.2). The increase in this metric after digtal transformation creates a clear business case for the pilot project.

A critical factor in digital transformation planning is recognizing the impact of sunk costs—previous investments in legacy systems that may no longer be viable. The plant, at eight years old, is likely to have a remaining operational life of 12–15 years. Any digital transformation investment should be evaluated against this timeframe, ensuring that the anticipated ROI aligns with the remaining lifespan of the plant.

Additionally, reducing technology debt—defined as the cost of maintaining outdated infrastructure—should be factored into the business case. By proactively upgrading to modern digital systems, enterprises can avoid escalating maintenance costs and operational inefficiencies that compound over time.

6.2.2 Digital Transformation Readiness Assessment

One of the first steps toward implementation is a gap analysis between the current state of the plant and the enterprise with respect to the digital transformation and the desired or

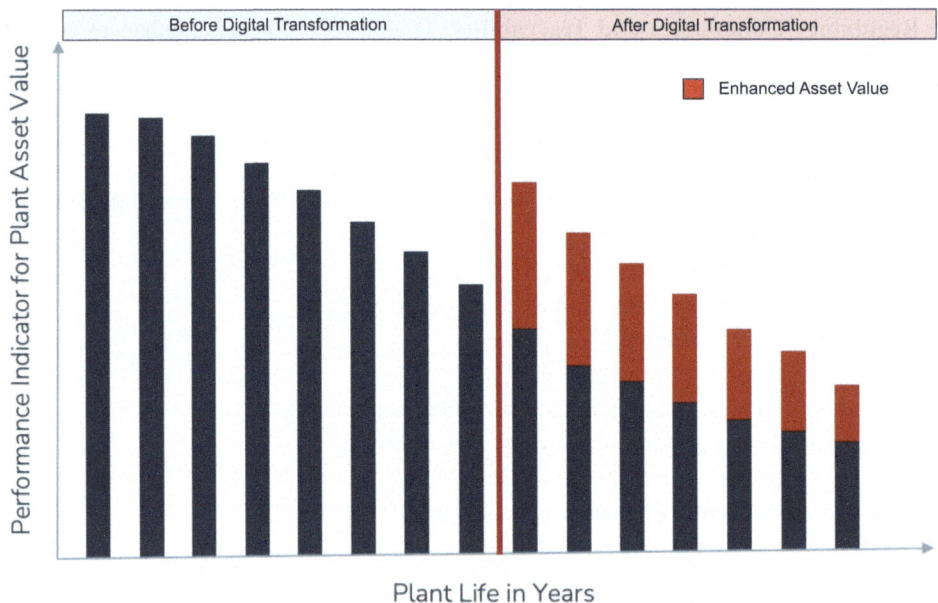

Fig. 6.1 Depreciating process plant assets can be retrofitted with digital transformation technologies to increase their value. Reduced OpEx after digitalization makes the plant more profitable over its residual life

anticipated fully digitalized state of the plant. This analysis provides a reasonable preliminary expectation of the transformation cost, effort, and resources needed. Here, we provide a simple methodology that can be readily developed by any digitalization implementation team or task force, and can be used consistently for different digitalization efforts within the same enterprise (such as multiple plants within the same organization).

The retrofit digitalization readiness assessment is designed to evaluate the current state of a plant's digital infrastructure and identify gaps relative to a fully digitally transformed state. The assessment covers six key categories, each with a total score of 5, leading to a final assessment score that can be visualized using a spider chart. For simplicity, the response to each question is assigned a binary value: **0**–No (feature not implemented) and **1**–Yes (feature implemented or available). Each category contains five questions, with scores normalized to a total of 5 for each category.

Table 6.1 provides an example of a digitalization readiness assessment questionnaire with the third column (response) filled in with a representative sample response that the digital transformation assessment team might obtain from their survey of the plant, the enterprise operational workflows and the workforce.

6.2 Pilot Implementation in a Mid-Life Manufacturing Unit

Table 6.1 Digitalization readiness assessment

Category	Question	Response (0/1)	Score
1. Process digitization	Are inline sensors connected to PLCs?	1	1.0
	Are most measurements done manually? (Inverse Score)	0	1.0
	Is process data stored digitally in historians?	1	1.0
	Is laboratory data recorded digitally?	0	1.0
	Are data entries automated without paper logs?	1	1.0
2. Automated process control	Are automated control loops implemented?	1	1.0
	Can PLC setpoints adjust field equipment in real-time?	1	1.0
	Are feedback loops used for process adjustments?	1	1.0
	Are closed-loop control strategies in place?	0	1.0
	Is there integration between field and control devices?	1	1.0
3. Enterprise level connectivity	Is SCADA data integrated with enterprise databases?	0	1.0
	Is there a CMMS system for asset tracking?	1	1.0
	Is a LIMS system in place for lab data management?	0	1.0
	Is ERP connected to process data sources?	0	1.0
	Do different enterprise systems interoperate?	1	1.0
4. Condition-based monitoring	Are condition-based monitoring tools used?	1	1.0
	Are there automated alerts for process anomalies?	1	1.0
	Is predictive maintenance implemented?	0	1.0
	Are AI/ML analytics used for monitoring?	0	1.0
	Are alarms linked to root cause diagnostics?	1	1.0
5. Workflow digitalization	Is decision support used in operations?	1	1.0
	Is decision automation applied for process control?	0	1.0
	Are operational workflows digitized?	1	1.0
	Are data-driven insights used for optimization?	1	1.0
	Is real-time transparency enabled across sites?	1	1.0
6. Workforce readiness	Are employees trained on digital tools?	1	1.0
	Are process engineers familiar with digital automation?	1	1.0
	Is there in-house expertise for digital twins?	0	1.0
	Are process models used for optimization?	0	1.0
	Is there a structured digital transformation roadmap?	1	1.0

The total score for each category is calculated by summing the individual scores, ensuring a maximum of 5 per category. The final assessment results can be visualized using a spider chart with six axes, representing each category. Table 6.2 shows aggregated scores for the six categories based on representative responses for two plants using the questionnaire in Table 6.1.

Table 6.2 Categories and scores for two plants (max.score/category: 5)

Category	Plant 1.	Plant 2
Process Digitization:	4	2
Automated Process Control:	4	2
Enterprise Wide Connectivity:	3	2
Condition-Based Monitoring:	2	1
Workflow Digitalization:	3	1
Workforce Readiness:	3	1

These values can then be plotted as a radar or spider chart to provide a visual representation of the plant's digitalization readiness. Figure 6.2 depicts the visual data for the two sample plants evaluated and scored in Table 6.2. It is evident that different process plants can be mapped on a chart to visually indicate how their pre-digitalization state differs from each other. Clearly, Plant 2 in the chart is less prepared for digital transformation than Plant 1 in Fig. 6.2 and it will take more resources and effort to digitally transform this plant. It is also interesting to note how much improvement can be achieved on each of the six categories

Fig. 6.2 Radar or Spider chart representation of the digitalization readiness scores of two plants evaluated using the same scoring methods shown in Table 6.1

of digital transformation. This type of an approach can provide guidance for adopting a reasonable pathway toward digital transformation.

For Plant 1, which has a higher level of process digitization and automated process control, the effort required to develop a comprehensive historian database will be minimal. Establishing such a database, along with software-based integration of a functional LIMS and CMMS system, can quickly lay the foundation for enterprise-wide connectivity. This would enable seamless data integration with a cloud-based or enterprise-level repository, along with efficient database querying and analysis capabilities. With some additional workforce training to implement condition-based monitoring, the plant could readily adopt a decision support system powered by advanced analytics. Since the necessary networking and software solutions would require only minimal modifications, the capital expenditure (CapEx) for this digital transformation project would be relatively low. The primary costs would stem from data storage, IT infrastructure, communication channels, and workflow digitization. Given these conditions, the expected benefits—such as improved key performance indicators (KPIs) and a strong return on investment (ROI)—could be realized in a short time frame.

Conversely, Plant 2 would require significantly more effort to digitize both its operational technology (OT) and information technology (IT) infrastructure. The lack of automated process control suggests a limited number of process monitoring sensors, which directly increases the cost of digital transformation. Installing conventional wired sensors within a distributed control system (DCS) framework can be costly and complex. Even if IIoT based sensors are considered, integrating its networking components with a legacy DCS network will require middleware and cybersecurity considerations. Additionally, the digitization process at the plant floor would be time-intensive and could potentially disrupt ongoing operations of the plant. Retrofitting the process control framework to achieve digital transformation readiness would be a more substantial and time-consuming undertaking for this plant, requiring careful planning to minimize operational disruptions.

6.2.3 Strategic Roadmap for Implementation

A phased implementation approach is recommended to minimize disruption and optimize investment efficiency:

1. **Phase 1: Data Infrastructure Upgrade**—Implement a SCADA historian, LIMS, and cloud/edge data processing capabilities.
2. **Phase 2: Automation and Advanced Analytics**—Introduce machine learning models for predictive maintenance, process optimization, and anomaly detection.
3. **Phase 3: Workforce Digital Enablement**—Provide training and upskilling to ensure smooth adoption of digital tools.
4. **Phase 4: Continuous Improvement and Optimization**—Establish a framework for ongoing system updates, recalibration, and cybersecurity enhancements.

By aligning digital transformation with economic goals, operational efficiency, and long-term plant sustainability, organizations can ensure a smooth and profitable transition to Industrie 4.0, paving the way for future technological advancements.

6.3 Workflow Automation: The Lowest Hanging Fruit in Digital Transformation

Workflow automation represents one of the most accessible and impactful steps in the digital transformation of process plants. By automating routine workflows, plants can significantly enhance efficiency, reduce human intervention, and expedite the conversion of raw data into actionable insights. This transformation aligns closely with the Data-Information-Knowledge-Wisdom (DIKW) framework, which guides the structured evolution of data into wisdom-driven decisions.

6.3.1 Workflow Automation and the DIKW Framework

In a traditional process plant, vast amounts of data are generated continuously. However, this data often remains underutilized due to its storage in disparate systems, reliance on manual logging, or lack of integration with higher-level analytical tools. The DIKW framework provides a structured approach to transforming this raw data into meaningful and actionable knowledge:

1. **Data:** The lowest level of the DIKW hierarchy consists of raw data points collected from sensors, log sheets, and control systems. Examples include temperature readings, pressure values, and flow rates.
2. **Information:** When structured and stored in a historian or an enterprise database, raw data gains context. For instance, tracking a temperature sensor over time allows trends to be observed, creating a time-series dataset.
3. **Knowledge:** By applying analytical algorithms or machine learning models, structured information becomes knowledge. For example, recognizing that a gradual temperature rise in a heat exchanger correlates with fouling allows proactive maintenance scheduling.
4. **Wisdom:** The highest level of DIKW is the application of knowledge to make intelligent decisions. A well-integrated decision support system (DSS) can autonomously detect anomalies, alert operators, and even recommend corrective actions, ensuring process stability and efficiency.

6.3.2 Examples of Workflow Automation in Process Plants

Workflow automation integrates plant data into a centralized information repository, which can then be autonomously queried at regular intervals by analytical software. This process accelerates the transition from data to knowledge, facilitating predictive decision-making and reducing response times. Some examples in process engineering include:

- **Condition-Based Maintenance:** Instead of relying on periodic manual inspections, an automated system continuously analyzes vibration and thermal data from rotating equipment. If abnormal trends indicative of misalignment or bearing wear are detected, the system automatically schedules maintenance before a failure occurs.
- **Automated Anomaly Detection:** A historian database continuously tracks pH fluctuations in a wastewater treatment process. When a significant deviation from baseline values is detected, an automated decision support system alerts the operators, suggesting corrective dosing of neutralizing agents before regulatory compliance issues arise.
- **Energy Optimization:** A process plant integrates real-time power consumption data with production rates. The system uses machine learning models to optimize compressor, blower and pump operation, reducing energy waste without affecting production.

6.3.3 Key Benefits of Workflow Automation

The primary advantage of workflow automation is the near-instantaneous awareness of process drifts and anomalies, compared to legacy systems where such deviations often remain undetected until significant performance deterioration occurs. Figure 6.3 schematically depicts the key benefits of workflow automation utilizing a decision support system. In traditional operations, identifying and resolving process issues follows a sequential and time-consuming path:

1. An operator notices an abnormality, such as a sudden drop in production efficiency.
2. The operations team retrieves historical data and investigates trends.
3. The root cause is identified, often requiring collaboration between multiple teams.
4. A corrective action is determined and implemented, causing additional delays.

With an automated decision support system, these time lags are eliminated as the digitalized plant is continuously acquiring data, processing the information and converting it into knowledge regarding the constantly evolving state of the plant. As soon as an anomaly is detected, the system synthesizes knowledge regarding the potential issue and recommends immediate corrective actions. This enables a rapid transition from data to wisdom, ensuring that operators can identify root cause and implement solutions without the extensive delays

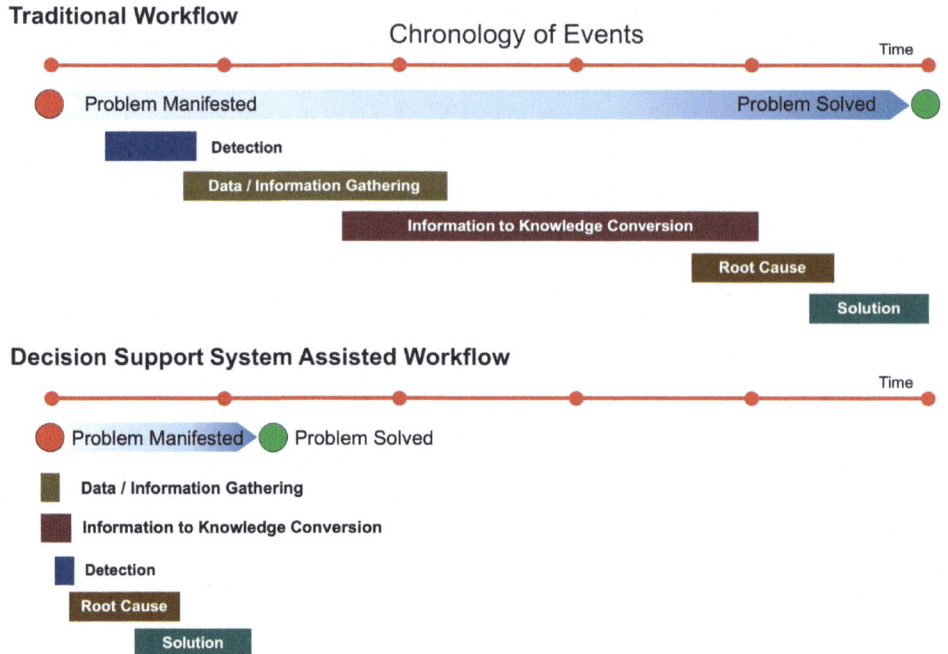

Fig. 6.3 Difference in workflows in traditional plant operation and an automated decision support system assisted plant. The automated and intelligent monitoring of plant information can lead to anomaly detection almost instantaneously. This allows a dramatic reduction in time required for diagnosing and troubleshooting a problem at the plant

inherent in manual troubleshooting. The result is a more resilient, adaptive, and efficient plant that can self-monitor its performance in real time, detect any anomalies as soon as conditions are met, and notify the pertinent operators about the problem.

In conclusion, workflow automation serves as a crucial entry point into digital transformation. By leveraging automation, structured data integration, and decision support systems, plants can minimize operational inefficiencies, enhance predictive capabilities, and significantly improve response times to process deviations. This transformation is not just an incremental improvement—it represents a fundamental shift toward a self-aware, data-driven, and intelligent manufacturing environment.

6.4 Enhancing Operations Through Transparency and Interoperability

Traditional chemical process plants consist of a sequence of unit operations, each performing a specialized function, such as, chemical reactions, separation, purification, heating, or cooling. In such plants, the output of one unit operation serves as the influent for the next in a

6.4 Enhancing Operations Through Transparency and Interoperability

continuous and interdependent manner. However, plant operators are often assigned responsibility for only specific unit operations, limiting their visibility into the performance of both upstream and downstream processes. This lack of transparency creates operational difficulty, as disturbances or suboptimal conditions in one part of the process may go unnoticed until they propagate downstream, leading to inefficiencies, reduced yield, or even equipment damage. By implementing digital transformation strategies that emphasize interoperability, plant-wide transparency can be significantly enhanced, allowing operators to proactively manage plant performance.

Interoperability ensures that data from different unit operations is aggregated into a centralized platform where real-time process variables, historical trends, and predictive insights are available for comprehensive analysis. When operators gain visibility into process conditions beyond their assigned units, they can anticipate potential issues before they escalate. For instance, let us consider the case of a seawater desalination plant through a process flow diagram in Fig. 6.4. In a desalination plant, if the multimedia filtration (MMF) unit experiences an increase in turbidity breakthrough, operators downstream at the reverse osmosis (RO) system can adjust pre-treatment chemical dosing or system recovery settings preemptively, mitigating potential fouling of the membranes and reducing unnecessary downtime for RO membrane cleaning. By leveraging interoperable digital systems, operators can access correlated process data through dashboards and analytics tools, enabling them to shift from reactive problem-solving to proactive optimization strategies.

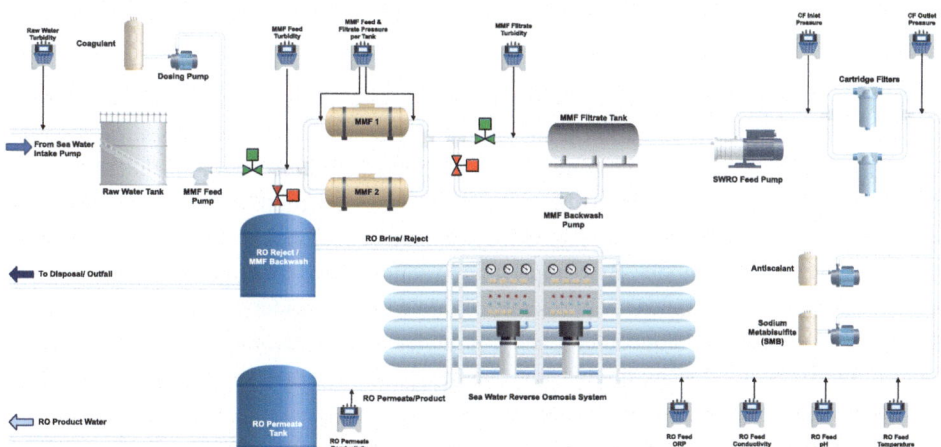

Fig. 6.4 A process flow diagram (PFD) of a seawater desalination plant, showing the sequence of different unit operations including chemical dosing, followed by multimedia filters (MMF), cartridge filter (CF), and reverse osmosis (RO) units. The PFD depicts the main in-line process monitoring instrumentation embedded in the system

6.4.1 Database Development and Analytics Integration

In the context of Industrie 4.0, transparency and interoperability are key pillars of a successful digital transformation. A seawater desalination plant consists of multiple unit operations, each with unique process parameters that must be continuously monitored and optimized. The primary goal of digital transformation is to implement a system that aggregates data from all these unit operations onto a centralized platform, ensuring a holistic view of the plant's performance. This section discusses the development of such a system and its impact on plant efficiency and decision-making.

To facilitate transparency and interoperability, a structured data aggregation system must be developed. This involves:

- **Centralized Data Repository:** A relational or time-series database that collects and stores process data from different unit operations, including chemical dosing, multimedia filtration, cartridge filtration, and the reverse osmosis (RO) system.
- **Data Acquisition:** Sensors and process control systems (PLC, SCADA) transmit real-time data on key operating conditions such as feed water turbidity, conductivity, pH, pressure, temperature, and flow rates.
- **Analytics Engines:** Algorithms and statistical tools to process historical and real-time data to detect trends, anomalies, and correlations.
- **Query Tools:** Web-based dashboards and visualization platforms allow operators to interact with data dynamically, making it possible to compare time-series plots and generate predictive insights.

Figure 6.5 presents a visualization of the key process status variables and inline sensor measurements for the desalination plant shown in Fig. 6.4. These variables are recorded at one-hour intervals over a 24-hour period. The overall plant state and the RO system status are represented as boolean values (0: Off, 1: On). The multimedia filter (MMF) unit operates in three states: 0 for Off, 1 for Filtration, and 2 for Backwash. All other variables are expressed as real numbers. By comparing the status of different units and the measured process variables at any given time, operators gain a clearer understanding of the dynamic interactions between interconnected plant components. This comprehensive data mapping over a unified time window highlights how changes in one unit operation impact downstream processes and overall plant performance.

This transformation redefines operational workflows in smart plants. As illustrated in Fig. 6.5, visualizing time series data of plant states (On/Off) alongside process measurements allows operators to identify and analyze sequential events within the plant. A live dashboard displaying these trends enhances plant-wide visibility, providing operators with a holistic view of process performance. Instead of relying solely on manual troubleshooting and reacting to process disruptions, operators can leverage data-driven decision support systems that detect potential anomalies in real time.

6.4 Enhancing Operations Through Transparency and Interoperability

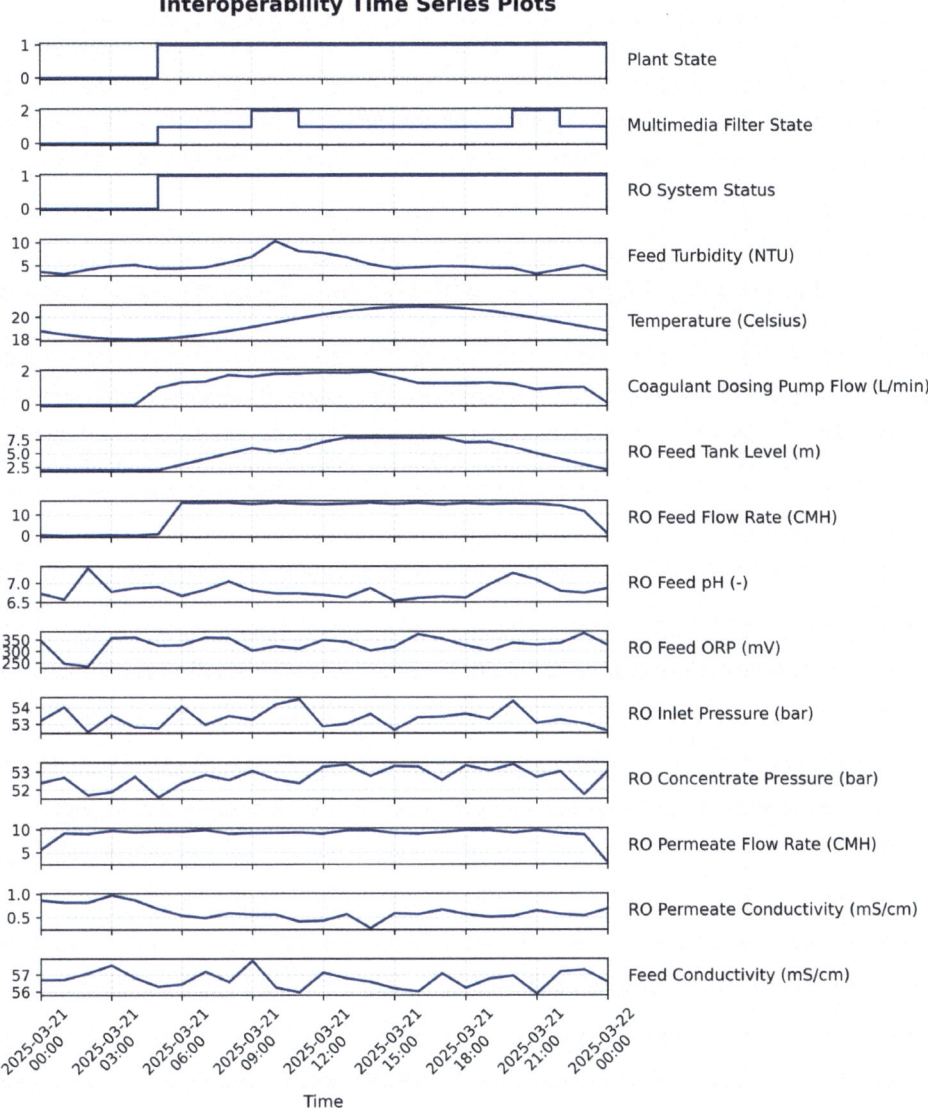

Fig. 6.5 Typical time series plots showing a transparent view of the states (On/Off) of different unit operations, as well as different process measurements for the seawater desalination plant of Fig. 6.4

Compared to traditional manual operations—where inefficiencies are often identified only after performance losses have occurred—a digitally transformed desalination plant enables immediate responses to changing conditions. By ensuring transparency and interoperability, operators gain continuous access to real-time plant performance data. This empowers them to make informed, data-driven decisions, optimize process efficiency, and enhance overall plant reliability.

6.4.2 Enhancing Plant Operations with Dynamic Recovery Optimization

One key advantage of digital transparency and interoperability is the ability to optimize plant performance dynamically. The desalination process is highly dependent on feed water quality, which fluctuates due to environmental conditions and seasonal variations. By implementing an intelligent analytics engine, the following operational improvements can be achieved:

- **Feedforward Control:** The system continuously monitors upstream influent quality parameters such as turbidity, salinity, and pH. If significant deviations are detected, pre-emptive adjustments can be made to chemical dosing rates and filtration settings.
- **RO System Adaptation:** Real-time monitoring of influent water quality enables dynamic adjustment of the RO system's recovery rate, preventing membrane fouling and optimizing energy consumption.
- **Predictive Maintenance:** Machine learning models analyze equipment performance over time, identifying early indicators of fouling or mechanical wear, reducing unplanned downtime.

6.4.3 Leveraging Transparency and Interoperability for Improved Performance

The digital transformation of the desalination plant offers several advantages over traditional manual operation:

- **Real-time Decision Support:** Operators receive immediate feedback on plant conditions and recommended actions, significantly reducing response time to process deviations.
- **Enhanced Operational Visibility:** A comparative dashboard integrates data from all unit operations, providing a unified view of plant performance.
- **Automated Process Adjustments:** By leveraging AI-driven insights, the plant can autonomously adapt its settings to optimize efficiency.

- **Reduced Human Error:** Automation and decision support tools minimize reliance on manual data entry and subjective judgment.

Transparency across unit operations ensures better coordination among teams, allowing plant personnel to communicate insights effectively and implement corrective actions in a timely manner. The result is an overall improvement in plant efficiency, resource utilization, and equipment longevity, ultimately leading to enhanced profitability and sustainability of operations. In essence, by adopting interoperability and transparency, traditional chemical process plants can evolve into smart plants where decision-making is streamlined, predictive maintenance is optimized, and plant-wide collaboration is fostered. This transformation marks a fundamental shift from reactive to proactive plant management, ensuring greater reliability, efficiency, and sustainability in seawater desalination operations.

6.5 Evolution and Maintenance of a Smart Infrastructure

In this chapter, we have outlined how a digital transformation journey can be initiated in any process plant, with a particular focus on retrofit digital transformation. We discussed the key preliminary steps, including developing a business case, setting clear goals, assessing the plant's baseline state, conducting a gap analysis, and implementing an initial digitization and digitalization framework. This initial phase aims to unlock some of the most accessible benefits of digital transformation. By applying these methods, proponents and stakeholders can gain firsthand insights into the value of transformation, paving the way for a more comprehensive, organization-wide implementation plan.

In the remainder of this chapter, we summarize the potential next steps in evolving the enterprise's digital infrastructure. We highlight the necessary improvements, outline strategies for scaling digital initiatives, discuss best practices for maintaining and optimizing the system, as well as the risks, barriers, and opportunities.

6.5.1 Connectivity and Integration: Building the Digital Thread

Once the basic framework of digitization is in place, particularly a viable approach of connecting the field level plant data collected through the OT layer into a consolidated database, the next step is to establish connectivity and integration across the manufacturing ecosystem. This involves linking disparate systems, machines, and processes to create a seamless flow of information. The establishment of a digital thread ensures that data is accessible and actionable at every stage of the production lifecycle. This connectivity sets the stage for real-time monitoring, analysis, and optimization.

6.5.2 Data Analytics: Extracting Actionable Insights

With a foundation of digitized and interconnected systems, manufacturers can leverage data analytics to extract meaningful insights. Advanced analytics, digital twins, machine learning, and artificial intelligence algorithms analyze the vast datasets generated by manufacturing processes. These insights enable informed decision-making, predictive maintenance, and process optimization, driving efficiency and cost-effectiveness.

6.5.3 Automation and Robotics: Enhancing Operational Efficiency

Automation and robotics play a crucial role in the digital transformation journey. By integrating intelligent machines into manufacturing processes, repetitive and labor-intensive tasks can be automated, reducing the risk of errors and increasing operational efficiency. This step involves the deployment of robotics for tasks such as assembly, packaging, and material handling, freeing up human resources for more complex responsibilities.

6.5.4 Human-in-the-Loop Automation: Augmenting Human Capabilities

Human-in-the-loop automation represents a symbiotic relationship between humans and machines. While machines handle routine and repetitive tasks, human operators remain integral for decision-making and complex problem-solving. This stage involves the integration of automation technologies that empower human operators, providing them with real-time data and decision support tools for more efficient and informed decision-making.

Decision Support Systems: Augmented Intelligence

As digital transformation progresses, decision support systems become pivotal. These systems leverage advanced analytics, AI, and machine learning to provide real-time insights and recommendations to human operators. Decision support tools empower operators to make faster and more accurate decisions, enhancing overall productivity and quality control.

Decision Automation: Towards Fully Autonomous Smart Plants

The final stage of the digital transformation journey involves decision automation. In fully autonomous smart plants, machines and systems make decisions independently based on predefined algorithms and learning from data. This level of automation reduces human intervention to strategic oversight and exception handling, allowing for increased operational efficiency, reduced response times, and enhanced adaptability to dynamic conditions.

6.6 The Challenges of Attaining a Fully Implemented Digitalization Vision

The prospect of fully closed-loop smart plants, driven by decision automation, holds immense promise for the manufacturing sector. However, the road to achieving this level of automation is fraught with strategic challenges that extend beyond technical complexities. From the intricacies of implementing sophisticated algorithms to addressing the human consequences, the journey towards fully autonomous smart plants requires careful consideration and strategic planning.

6.6.1 Overcoming Complexity and Integration Challenges

Implementing decision automation at the level required for fully closed-loop smart plants involves overcoming substantial technical difficulties. Integration of complex algorithms, machine learning models, and AI systems into existing manufacturing processes poses a significant challenge. Ensuring seamless connectivity and interoperability across a diverse range of systems and machines demands robust technical solutions and careful planning.

6.6.2 Data Quality and Reliability: A Cornerstone for Decision Automation

Decision automation relies heavily on data quality and reliability. Inaccurate or incomplete data can lead to flawed decision-making, potentially compromising the efficiency and safety of manufacturing processes. Establishing data governance frameworks and investing in data quality assurance measures are critical steps in addressing this challenge.

6.6.3 Cybersecurity: Safeguarding Against Threats

The increased connectivity inherent in fully closed-loop smart plants introduces new cybersecurity challenges. Protecting against cyber threats and ensuring the integrity of data and decision-making processes become paramount. Robust cybersecurity measures, including encryption, secure access controls, and continuous monitoring, are crucial to safeguard against potential cyber-attacks.

6.6.4 Job Displacement: Addressing Workforce Concerns

A significant strategic challenge in the journey towards fully autonomous smart plants is the potential impact on the workforce. Increased automation can lead to job displacement, particularly for roles that involve repetitive and routine tasks. Addressing these concerns requires a comprehensive approach that includes reskilling and upskilling initiatives to prepare the workforce for roles that complement and collaborate with automated systems.

6.6.5 Human Consequences: Balancing Efficiency with Ethical Considerations

As decision automation becomes more prevalent, ethical considerations come to the forefront. Balancing the quest for operational efficiency with the ethical treatment of the workforce is a delicate challenge. Manufacturers must consider the potential consequences of automation on job security, mental health, and the overall well-being of employees. This necessitates transparent communication, empathy, and a commitment to ethical business practices.

6.6.6 Reskilling and Retraining: Investing in Human Capital

To mitigate the impact of job displacement, manufacturers must proactively invest in reskilling and retraining programs. Offering opportunities for employees to acquire new skills and transition into roles that require human expertise, such as overseeing and optimizing automated systems, is essential. Collaboration with educational institutions and government initiatives can facilitate a smoother transition for the workforce.

6.6.7 Change Management: Cultivating a Culture of Adaptability

Implementing decision automation in fully closed-loop smart plants requires a cultural shift within organizations. Change management becomes a strategic imperative to cultivate a culture of adaptability, continuous learning, and collaboration between humans and machines. Fostering a positive attitude towards technological change and providing adequate support structures are crucial components of successful change management.

Integrating Digitalization into the Process Industry

7.1 Introduction

The implementation of digital technologies in the chemical process industry presents unique challenges and opportunities due to the inherent complexities of chemical processes. Unlike simpler manufacturing systems, chemical processes are governed by intricate thermodynamic, kinetic, and transport phenomena, making their digitalization a highly specialized endeavor. This chapter delves into the underlying concepts and technical nuances involved in embedding digital technologies within the chemical industry, providing a balanced perspective between purely data-driven analysis and physics-based modeling approaches.

A key aspect of digital transformation in the chemical process industry is the selection and deployment of appropriate sensors for process monitoring. The choice of sensors impacts the accuracy, reliability, and granularity of data collection. This chapter explores different types of sensors, their working principles, and their suitability for various chemical processes. Furthermore, data collection methodologies must be aligned with the kinetics and dynamics of chemical reactions and phase transitions to ensure meaningful and actionable insights. The timescales of data acquisition are crucial in capturing transient phenomena and steady-state operations alike.

Another critical consideration in digitalizing chemical processes is the analysis of collected data. This requires an understanding of the thermodynamic basis governing phase changes and reaction systems. By integrating thermodynamic principles, data interpretation becomes more robust and reflective of the underlying physical and chemical behaviors. Additionally, process modeling plays a significant role in digitalization, necessitating the application of conservation laws of mass and energy to develop predictive models that enhance process efficiency and optimization.

To manage and interpret large volumes of process data effectively, data dimensionality reduction techniques are often employed. This chapter discusses how traditional chemical process dimensional analysis can aid in identifying key process variables and reducing data complexity without compromising essential information. Such methods help in bridging the gap between data-centric machine learning approaches and first-principles physics-based modeling, ultimately leading to a hybrid approach that leverages the strengths of both paradigms.

This chapter is structured as follows: Sect. 7.2 covers sensor types, their selection criteria, and key factors to consider in their use and maintenance for the process industry implementations. Section 7.3 discusses data acquisition and conversion to information in alignment with process kinetics and dynamics. This section explores data interpretation techniques with a focus on thermodynamic and thermochemistry principles, outlining process modeling approaches based on mass and energy conservation principles.This leads to the development of physics based digital twins as a basis of processing the plant information and managing the plant operations. Section 7.4 examines data dimensionality reduction and its relevance to chemical process analysis. Finally, in Sect. 7.5, we provide an integrative perspective on combining data-driven and model-based approaches for process optimization.

Through this discussion, we aim to equip engineers and researchers with a comprehensive understanding of the digitalization of chemical processes, highlighting the potential benefits of concepts like process modeling, digital twinning, handling data with emphasis to process engineering embedded knowledge bases, as well as the importance of a hybrid analytical framework balancing data-driven and physics driven process digitalization.

7.2 From Stimulus to Data: Principles of Transduction

We cannot envision digital transformation unless we have a firm understanding of data. In the digital world, data represents the underlying common currency through which a myriad of information are collected, stored, processed, and interpreted by computational algorithms, assisting us with making decisions, and taking actions. Whether it is engineering, commerce, social sciences, humanities, economics, entertainment, or health, every practitioner has to wade through data to process information. In this respect, data can be seen as a great unifier of diverse information, knowledge, thoughts, and ideas. At some fundamental level of abstraction, our perception of hotness or coldness through temperature, speed of our car, savings in our banks, heart rates, listening to music, and watching a video, are all relegated to the flow of streams of binary bits–strings of zeros and ones–stored in memory blocks of computers in a single type of binary format.

Nearly any profession today involves use of data in some form. Data has become the lifeblood of the modern digital age, fundamentally transforming how businesses, governments, and individuals operate. The digital revolution has ushered in an era where data is generated at an unprecedented rate from various sources, such as, social media, sensors, transactions, and more. This data explosion is characterized by the three V's: volume, velocity, and variety. Volume refers to the vast amounts of data generated; velocity indi-

7.2 From Stimulus to Data: Principles of Transduction

cates the speed at which new data is created and processed; and variety encompasses the different types of data, including numbers, text, images, and videos. Understanding these characteristics is crucial for organizations aiming to harness data for competitive advantage.

In this chapter, we will first focus on how data is acquired from the physical plant and its environment, carefully studying the diverse types of sensors and transducers that measure and quantify the measured variables in chemical processing plants. Sensors and transducers are the devices that bridge the physical and digital realms. We will then take a look at the conversion of these sensor signals into bits and bytes that form digitized information. We will next discuss how the data is stored, retrieved and queried efficiently, and finally, how to ensure the integrity of the data pipeline. Data is the foundation of digital transformation, and ensuring the integrity and accuracy of the data pipeline is perhaps the most critical aspect of maintenance in digitized plants. This requires focusing on sensor maintenance, maintenance of data pipelines, ensuring data security, as well as accessibility to data on demand. In this context, how databases are maintained, as well as the evolving ramifications of so-called "big data", "data lakes" or "data oceans" will be discussed. Finally, we will close this chapter with a brief overview of what role data will play in future process plants when we build them with the "digital first" mindset.

7.2.1 Field Instruments

Field instruments are the essential sensory and motor connections of a process plant automation and control framework to the physical world. Field instruments are electromechanical or pneumatic systems performing two types of functions, namely, sensory elements, and motor or transduction elements.

The sensory elements convert a physical stimulus such as pressure or temperature into a measured force, strain, deflection, deformation or an electromagnetic induced field. These equipment are the sensory terminal points of cyber-physical systems, converting information about the plant equipment, the state of materials being processed in such equipment, the operating conditions, as well as the environment, into data for the digital control system. The basic mechanisms of such transducers is to convert the external stimulus to a measured output variable, which is eventually converted to an analog or digital electrical signal. Digitization of the physical information into predominantly binary data bits forms the first essential basis of digital transformation, which we refer to as digitization.

Another subset of field instruments, the motor elements, provide outputs from the digital control system to the physical plant, for example, opening and closing flow valves, or turning on- or off pumps and motors. Mechanical transducers are devices that convert a digital or analog electrical signal into an external force that create physical motion or action.

The translation mechanisms of physical stimuli perceived by sensors to the internal recording to an electrical output are diverse, and involve a variety of physical and chemical interactions with materials having special properties. These include, for instance, interaction of heat (temperature) with bimetallic strips to produce mechanical strain due to bending of the strip. The mechanism of transduction of heat to electrical signals can also be done using thermoelectric materials (such, as thermocouples) which produce an electric potential gradient as output when subject to temperature change. Optoelectronic mechanisms convert light to electrical signals, mechatronic devices can convert mechanical forces or strains into electrical or magnetic outputs.

Notwithstanding the mechanism of the front-end transduction, the back-end mechanisms of most modern field equipment are fairly standard. The older analog conversion mechanisms involved manipulation of a bridge circuit with pre-calibrated resistances to generate an output voltage in the range of 0 to 10 V or an output current in the range of 0 to 20 mA. More modern equipment consist of an additional analog to digital (A/D) converter that can scale the analog voltage or current into an integer number depending on the bit resolution of the A/D converter. For instance, a 16 bit A/D converter can convert an input analog signal of 0 to 20 mA into integers between 0 to 65,535 (unsigned) or −32,768 to 32,767 (signed). Most digital control systems can easily process these types of integer data.

For motor or transducer systems, the direction of data transmission is reversed, where the process control system provides the output signal in the form of binary states of bits, or as integers (8 bit or 16 bit words). The binaries are converted using digital to analog (D/A) converters to provide an analog electric signal to the transducer circuit. This electric signal (either the applied potential difference or the driving current) triggers the physical action at the output of the transducer. A simple example of a mechanical transducer is a mechanical relay, which is simply a switch that turns on or off based on the state of the input voltage. When the input voltage sent by the control system is high, the relay will turn on, whereas, the input voltage is zero, the relay will be turned off. More complicated actuation systems involve motor speed controls that are operated using variable frequency drives using pulse width modulation or other digital circuitry principles.

7.2.2 Sensors for Process Monitoring

Data is generated when we aim to process information through measurements. Measurement is a technique whereby we process information related to the physical world through an appropriate sensor (sight, sound, smell, taste, temperature, etc.). The sensor is generally able to characterize and classify (or bin) the information based on some characteristic, such as the intensity or frequency of a sound, the brightness (intensity) or color (wavelength) of light, and the degree of hotness or coldness (temperature). Sensors are devices that quantize information. They typically have limits of measurement, within which they can calibrate and quantitatively differentiate between different levels of the measured information.

For instance, a thermometer can have a lower limit of measurement as freezing point of water (zero degree Celsius) and boiling point of water (100 degrees Celsius) as the upper limit of measurement. This defines the range of the sensor (from zero to 100 degrees

Celsius). When designing such a thermometer, the designer can graduate the thermometer to equally divide the 100 degree range into eleven divisions, where each 10 degree increment is marked (as 0, 10, 20, ..., 100). This represents a scale, where the temperature input can be measured at a resolution of 10 degrees. Using this thermometer, we can quantify the temperature of any physical environment and place these into different bins. For instance, if this thermometer is placed inside a refrigerator, it's reading will fall in a bin between 0 and 10 degrees Celsius. When in a normal air conditioned room, the thermometer may record a temperature between 20 and 30 °C. With this thermometer, we can quantize our perception of temperature of various environments with a resolution of 10 °C.

If we place this thermometer in an oven and we note that the reading is in a bin between 20 and 30 °C, we typically *infer* that the oven is in an "off" state, whereas if the thermometer shows a value of 100 (maximum of the scale) we will conclude that the oven is in an "on" state, and is heated up. One challenge of this thermometer is its scale limit. When the temperature indication is 100 °C, it does not necessarily imply that the temperature of the oven is 100 °C. It is rather more likely that the temperature of the oven is higher than the maximum limit of the thermometer.

The example of the thermometer discussed above is the typical realm of quantizing information through analog sensing. This is the fundamental premise of analog instrumentation. Most instrumentation for physical variables utilize some form of stimulus from the environment that stimulates a change in the sensor's internal mechanism (transduction). For example, classical thermometers (liquid expansion type) utilized the thermal expansion or contraction of liquids in a narrow capillary tube. Another class of thermometers use bending of bimetallic strips due to heat. Yet another class of thermometers use flow of electrons engendered by heat between dissimilar metals (thermocouples). The first step toward measurement is the design of the analog transducer that can accurately measure the physical quantity.

Sensors are used in the chemical manufacturing industry to measure conditions inside bulk fluids or solids, often inside sealed tanks that are pressurized, containing chemicals at extremes of temperatures. Pressure and temperature are two primary variables that are ubiquitously measured in most processes. Another important process parameter that requires measurement is flow rate, which can be mass or volumetric flow rates of streams. Finally, depending on the process, a variety of measurements are performed to quantify the chemical composition, fluid properties, and quality related parameters of the fluid streams. This renders process monitoring in the chemical processing industry to be rather diverse compared to many other sectors of the manufacturing industry.

In mechanical or solids manufacturing environments, the measurement and sensory technique that often suffices for process and product quality monitoring is visual observation, which is achieved through cameras. This makes vision based systems the most ubiquitous data acquisition method in such manufacturing environments. One can perform counting of units, finding defects in features of individual units manufactured, as well as sorting of solid objects in processing plants entirely through vision based systems. Robotics and automation in automotive, semiconductor, bulk sizing and packaging of solid products manufacturing are mostly driven by vision based techniques.

In contrast, the process monitoring systems in chemical processing plants must probe the states of fluids (liquids, gases, as well as fluidized solids or powders) in confined environments of pipes and vessels. Visual inspection of these through camera is almost never translated into quantifiable data, therefore requiring introduction of different approaches of transduction of the observation into a quantized variable. For instance, a common type of measurement can be that of electrical conductivity of the liquid inside a pipe. This is performed through a pair of electrodes dipped in the liquid. Similar measurements in aqueous systems using electrodes include measurements of pH, oxidation reduction potential (ORP), specific ion concentrations, *etc*. Many of these measurements are electrochemical.

Apart from the four fundamental types of measurements (temperature, pressure, composition, and flow), there are a variety of other measurements required in a chemical manufacturing plant, such as volume, tank level, mixing or stirring rate, conversion in a reactor, *etc.*. These are often derived from the plant unit operation or process design and the four fundamental measurements representing the flow rates and states of the substances flowing in the plant.

Table 7.1 outlines the mechanisms of sensing typical process parameters in a chemical processing plant, including transduction methods for pressure, temperature, flow, and composition.

Electrochemical transducers utilize small voltages and currents generated from the interaction of suitable transducers with the charged entities in the liquid. These electrochemical sensors must be in contact with the fluid, and hence, must be inserted into the pipes or tanks. Some chemical measurements are performed spectroscopically, which entails the interaction of light beams with the fluid medium. While these methods are optical, they are fundamentally different from imaging using cameras. An important aspect of such methods are that the source of light must also be introduced into the liquid stream to activate the sensors and transducers. For instance, in submerged ultraviolet (UV) absorption or light scattering based measurements of chemical concentrations or turbidity in aqueous streams, light sources of appropriate wavelengths paired with a detector are introduced to the liquid stream as a combined package with a pre-defined distance (path length) between the light source and the detector. As the fluid passes through the gap between these two units, the chemical species in the stream interfere with the light beam, affecting the transmission of light, which is quantitatively captured by the detector. The detectors are typically photoelectric or charge coupled devices that respond to the incident light by generating an electrical effect that can be measured as a change in voltage or current in a secondary circuit.

It is therefore discernible that sensing and measurement of conditions in chemical processes, and acquisition of information in the form of data is quite diverse in the chemical industry. A critical aspect of these devices is that they must be embedded into the fluid streams, and can be affected by the fluids. For instance, the fluid can contain particles or impurities that deposit on the electrodes, detectors, or obstruct the light sources, causing the

7.2 From Stimulus to Data: Principles of Transduction

Table 7.1 Different types of sensors used in the chemical manufacutring environments

Process parameter	Sensing mechanism	Transduction principle	Details
Pressure	Strain gauge	Resistive	Measures strain on a diaphragm caused by pressure, changing resistance is used to calculate pressure
	Capacitive sensor	Capacitive	Pressure changes the distance between capacitor plates, altering capacitance and generating a pressure reading
	Piezoelectric sensor	Piezoelectric	Pressure causes deformation of piezoelectric crystals, generating a voltage proportional to the applied pressure
	Bourdon tube	Mechanical	Tube deforms under pressure, and mechanical movement is converted into a pressure reading via a dial or gauge
Temperature	Thermocouple	Seebeck Effect	Two dissimilar metals generate a voltage based on the temperature difference between junctions
	Resistance temperature detector (RTD)	Resistive	Resistance of a metal (typically platinum) changes with temperature, providing a direct temperature reading
	Thermistor	Resistive	Resistance of semiconductor materials changes non-linearly with temperature
	Infrared sensor	Radiative	Measures the infrared radiation emitted by an object to determine its temperature
Flow	Orifice plate	Pressure Drop	Flow through a restricted orifice generates a pressure differential, which is used to calculate flow rate
	Turbine flow meter	Mechanical rotation	Flow causes a turbine to rotate; the rotational speed is proportional to the flow rate
	Electromagnetic flow meter	Faraday's Law of Induction	Conductive fluid flowing through a magnetic field induces a voltage proportional to the flow velocity
	Ultrasonic flow meter	Doppler Effect / Transit Time	Measures the time it takes for an ultrasonic signal to travel between transducers, with flow velocity altering the time or frequency shift
Composition	Electrochemical sensor	Chemical Reaction (Electrochemical)	Detects specific gases or ions by measuring current generated by redox reactions at an electrode (e.g., oxygen sensor, pH sensor)
	Ion-selective electrode (ISE)	Selective Ion Exchange (Electrochemical)	Measures concentration of specific ions by monitoring the potential difference across a selective membrane
	Optical absorption (UV/Visible/IR)	Absorption Spectroscopy	Measures the absorption of specific wavelengths of light by a sample to determine chemical composition
	Raman spectroscopy	Inelastic Scattering (Optical)	Measures shifts in the frequency of scattered light to identify molecular bonds and composition
	Fluorescence spectroscopy	Excitation and Emission (Optical)	Detects specific chemical species by exciting the sample with light and measuring emitted fluorescence
	Fourier-transform infrared (FTIR)	Interferometry (Optical)	Measures how a sample absorbs infrared light at different wavelengths to determine composition
	Electroacoustic sensor	Acoustic Impedance	Measures changes in acoustic wave properties (e.g., velocity or attenuation) as they pass through a sample, influenced by chemical composition
	Tunable diode laser absorption spectroscopy (TDLAS)	Absorption Spectroscopy (Optical)	Uses a tunable laser to measure gas concentrations by monitoring absorption at specific wavelengths
	Potentiometric gas sensor	Nernst Potential (Electrochemical)	Detects gases like CO_2 or O_2 based on the potential difference generated between a reference electrode and a gas-sensitive electrode
	Conductometric sensor	Electrical Conductivity (Electrochemical)	Measures changes in conductivity due to the presence of specific ions or chemicals in the sample

measurements to become erroneous. This is referred to as the fouling of sensors, causing reversible or irreversible drifts in the measurement. This type of fouling is often not encountered in vision based systems during solids processing as the cameras are not susceptible to fouling by the measured objects.

Measurement of flow is another important aspect of monitoring process conditions and throughput in a chemical plant. Flow measurement in liquids is also significantly more challenging than vision or echolocation utilizing light, ultrasonic, or electromagnetic waves (LIDAR or RADAR) based remote sensing of rigid object motion (used in robotics and autonomous vehicles). Liquid flow rates are often measured directly by immersing the measurement device into the streams. Mechanisms like rotation of a turbine, magnetic field fluctuation, sound speed or frequency change, as well as pressure drops across specially designed pipe fittings (venturi or orifice) are commonly used as direct techniques of flow measurement. All of these methods are susceptible to deviations or drift due to the change in composition of the liquid or fouling by contaminants in the liquid. Indirect or remote sensing methods that probe flow velocities sometimes use sound waves, and can be mounted outside pipes, but are susceptible to several external conditions that affect their accuracy. Many large scale gas flows are often measured using hot wire anemometry, which involves tracing heat transfer from a hot wire immersed in a flowing gas (as long as the gas in not flammable). Ideally, mass flows are the most rigorous types of flow measurements that can be performed on chemical systems as liquid and gas densities vary with temperature and pressure. However, mass flow measurements are often quite complex, leading to more conventional volume flow measurement devices.

Instrumentation for measurement of chemical species compositions in a mixture are of diverse types, and depending on the process complexity and sophistication, as well as product value, in-line instrumentation for these measurements can range from simple pH or solution conductivity measurements, which are electrochemical devices, specific ion concentration measurement devices, which are colorimetric, suspended solid concentration measurements using light scattering, transmission, or absorption, or electroacoustic mechanisms in concentrated solid dispersions or suspensions. In some complex applications, devices such as chromatographs, or mass spectrometers are embedded in a plant for in-line composition monitoring. These instruments take a slip-stream from the plant or process vessel, and perform analysis on the stream at regular time intervals.

In summary, the first step of converting a physical plant information into digital data is to convert the measured property (input signal) from the physical process into an output signal, whether it is pressure, strain, rotational speed, electrical voltage or current, a magnetic field intensity, light, or heat. The quantized output signal forms the classical analog output, the change of which indicates the change in the input signal. For this indication to be quantitative and reliable, the transducer mechanism should accurately and repeatedly provide the same output signal for the same input signal intensity. It should also ideally change the output signal intensity linearly in response to the change in input signal. The linearity of scaling is often assumed in most sensors within its specified range, and this scaling parameter between the input and output signal is referred to as the sensitivity of the sensor.

Basic Sensory Transduction–Linear Sensors

A typical equation relating the input and output signal for a sensor is often a linear or nonlinear function that describes how the sensor converts a physical quantity (input) into an output signal (for example a voltage). For many sensors, the relationship can be approximated by a linear equation:

$$V_{\text{out}} = S \cdot Q_{\text{in}} + V_{\text{offset}}$$

where V_{out} is the output voltage (or signal) from the sensor, S is the sensitivity of the sensor, which indicates how much the output signal changes per unit of input quantity, Q_{in} is the input quantity being measured (*e.g.*, temperature, pressure, displacement), and V_{offset} is the offset voltage, representing the sensor's output when the input quantity is "zero" (or at the lowest end of the range of the sensor).

For example, in a temperature sensor (thermocouple), if the sensor's sensitivity (S) is 10 mV/°C, and the offset voltage (V_{offset}) is 0 mV, the output voltage (V_{out}) can be calculated as:

$$V_{\text{out}} = 10\,\text{mV}/^\circ\text{C} \cdot \text{Temperature}\,(^\circ\text{C})$$

In some cases, the relationship may be nonlinear and could involve higher-order terms or logarithmic functions, especially for sensors like thermistors or photodiodes. A more general relationship for a nonlinear sensor could be:

$$V_{\text{out}} = f(Q_{\text{in}}) + V_{\text{offset}}$$

where $f(Q_{\text{in}})$ represents a nonlinear function describing the sensor's behavior. The exact form of f depends on the sensor's characteristics and how it responds to the input quantity.

Timescales of Sensing

So far in our discussion of analog sensors, we have neglected time by implicitly assuming that the process variable (input) being measured is instantaneously translated to the output signal through a transducer. This is often not the case. There is a finite time required for the sensor to respond to a change in the process conditions. The time lag depends on the mechanism of the sensor. For some variables, the response of the sensor could be almost instantaneous. However, in many chemical processing situations, the interaction between the fluid and sensor evolves more slowly, and these transient stages when the sensor is changing it's output can be quite prolonged. for instance, in electrochemical measurements, the kinetics of reaction of charged species with the electrodes can be quite slow (order of seconds), whereby if the fluid conditions (input signal) undergo a step change, the output signal takes a few seconds to register a corresponding change. Within these few seconds, the output signal changes following some type of kinetic response to the input signal. This delay should be accounted for, and if the input conditions change more rapidly than this sensor response delay time, the measurements can either be quite erroneous, or these may require considerable attention from experts to ensure correct interpretation of the sensitivity or offset of the sensor output signal. The dynamic response of sensors are therefore quite important in assessing the accuracy of measurements.

The dynamic response time of sensors imposes limits on the frequency of data acquisition for time series data. The time interval for the dynamic response of sensors to a measured variable change is typically represented by the sensor's "time constant", τ. The time constant is a measure of the time it takes for the sensor to respond to a step change in the input signal and reach about 63.2% of its final value.

A simple first-order linear differential equation can represent the dynamic response of many sensors. This is given by:

$$\tau \frac{dV_{out}(t)}{dt} + V_{out}(t) = K \cdot Q_{in}(t)$$

where, $V_{out}(t)$ is the output signal at time t, $Q_{in}(t)$ is the input quantity (measured variable) at time t. K is a proportionality constant (gain), and τ is the time constant of the sensor.

For a step change in the input quantity from $Q_{in,0}$ to $Q_{in,\infty}$, the output signal $V_{out}(t)$ over time can be expressed as:

$$V_{out}(t) = V_{out,\infty} + (V_{out,0} - V_{out,\infty}) \cdot \exp\left(-\frac{t}{\tau}\right)$$

where, $V_{out,0}$ is the initial output signal at $t = 0$ and $V_{out,\infty}$ is the final output signal as t approaches infinity.

This equation shows that the output signal $V_{out}(t)$ exponentially approaches the final value $V_{out,\infty}$ with a characteristic time constant τ. After a time interval of about 5τ, the sensor output is typically considered to have settled to within less than 1% of the final value, effectively representing the sensor's dynamic response time.

This brings us to a real world problem with many types of sensors, namely, what happens when the signal captured by the sensor changes more rapidly than this characteristic dynamic response time of 5τ of the sensor to acquire a stable single reading? What is a good data acquisition interval for such a sensor? What is the accuracy and reliability of data collected too frequently *vs.* less frequently?

In classical time series data analysis there exists a very well known theorem in signal processing called the *sampling theorem* that provides a reasonable approach for ensuring how frequently the data should be acquired in time such that all features of the time dependent signal are retained [1]. An example of the application of the sampling theorem exists in the frequency of sampling of audio or video data, and audio or video compression algorithms. The sampling theorem simply states that if a continuous time series waveform (periodic data) has a characteristic frequency of n Hz, then a discrete signal represented by minimum sampling frequency of $2n$ Hz will be necessary to record the features of this waveform accurately. If the sampling frequency is lower, then the acquired discrete information will provide a distorted and inferior quality representation of the actual signal.

Sensor Qualities

The "five R's of sensors" are crucial principles that ensure the reliability and accuracy of sensor data, fundamental for applications in various fields such as industrial automation, healthcare, and environmental monitoring. These principles include Repeatability, Reproducibility, Resolution, Response Time, and Range. *Repeatability* refers to the sensor's ability to consistently produce the same results under unchanged conditions. *Reproducibility* is the capacity of the sensor to yield the same results when measurements are taken by different users or equipment. *Resolution* defines the smallest change in a physical quantity that the sensor can detect. *Response Time* is the speed at which the sensor reacts to a change in the measured variable. Lastly, *Range* is the span between the minimum and maximum values that the sensor can accurately measure. Together, these characteristics ensure that sensors provide precise, reliable, and meaningful data for various applications.

7.2.3 Workflow and Business Considerations for Sensor Integration

Frequently, process industry digital transformation task force members and plant owners tend to believe that digitalization implies adding sensors and linking them up as internet of things (IoT) objects. This leads them to add sensors of every conceivable type to their plant without properly accounting for their relevance to the process and their benefits. While additional sensors do put more information at our disposal, there are several problems that immediately arise:

1. Operators need to be trained to use and maintain the new sensors,
2. There is no clear understanding of how the sensor adds overall value to the process, and how it enhances the overall utilization of information for profitability,
3. When the sensors do not provide the desired benefits, it leads to incorrect and inadequate monitoring of the plant, leading to the notion that digitalization has no value.

The key lesson here is that sensors and IoT are just one small component of digital transformation. Simply adding sensors to a plant or a process cannot lead to an overall attainment of the benefits of digitalization. It is important to understand what maximizes the benefits of digitalization, how each additional automation component injects efficiency into the workflow, and take steps that are the "lowest hanging fruits" to attain a successful digital strategy implementation.

7.3 From Data to Process Information

The advancement of digital technologies in process engineering has enabled real-time data acquisition and analysis, fostering better decision-making and predictive capabilities. This section presents a structured framework for extracting vital process information from plant data, and knowledge synthesis in chemical process plants. The approach integrates mapping sensor data to a set of process information utilizing thermodynamic property estimation, transport phenomena modeling, reaction kinetics, and network-based plant modeling. Additionally, we introduce the use of dimensionless numbers for data-driven analysis and digital twin development.

7.3.1 Thermodynamic Information

Accurate process monitoring starts with capturing sensor input data that measures key process variables such as **pressure, temperature, flow rates, and composition** at critical process points and for all relevant process streams. These measurements serve as inputs to thermodynamic models, which estimate state properties (e.g., enthalpy, entropy, and specific volume) [2] and transport properties (e.g., viscosity, thermal conductivity, and diffusivity) [3].

Fundamentally, pressure and temperature are two critical parameters that form the backbone of all measurements in a chemical processing plant, providing considerable information about the materials being processed in the plant. These two parameters (two fundamental state variables) allow assessing and predicting the state of most common materials encountered in a chemical process plant. For instance, if we consider a tank containing pure water at a temperature of 120 °C and a pressure of 5 bars, we can determine the state (liquid or gaseous), density, internal energy, enthalpy, specific heat, entropy, exergy, thermal conductivity, viscosity, *etc.* of the water [4, 5]. Furthermore, with a little thermodynamic insight, we can easily discern what will happen if we release the pressure of the tank to 1 bar. This can be entirely simulated *in-silico*, without actually releasing the pressure in the physical tank, utilizing thermodynamic equations of state such as the Peng-Robinson or Redlich-Kwong equations, and empirical or theoretical property estimation methods (e.g., corresponding states principle or group contribution methods) [4]. The thermodynamic insight is a critical piece of knowledge that chemical engineers and chemists acquire about the relationship between temperature, pressure, and physical or chemical properties of a substance. These thermodynamic calculations form the basis of design of chemical processing plants, and are critical in analysis of the process performance.

Generally, for a multi-component system (for example, a mixture of sugar and water, or an aqueous salt solution), knowing the temperature, pressure, and mixture composition can fix the state properties of the mixture. The measurement of composition in a multicomponent chemical mixture can be a little more involved. Composition of a mixture is determined by

7.3 From Data to Process Information

measuring the concentration of different components (chemical species) of interest present in the mixture. Ideally, for an n-component mixture, independent measurement of $(n-1)$ component concentrations will be necessary to unequivocally specify the composition of the mixture. In many cases, several simplifying assumptions can be made about the mixtures (such as dilute solutions), where one of the components is present in an overwhelmingly large quantity compared to the other constituents. Such assumptions can provide several simplified pathways to ascertain the composition of these types of mixtures [2].

A multicomponent chemical stream consists of various species. For instance, air is a mixture of gases with oxygen and nitrogen being the dominant components along with smaller amounts of CO_2, Ar, and water vapor (H_2O) among a multitude of other species in trace quantities. The species refer to molecules of different types that are present in a multicomponent system, whereas the term composition is defined as a fraction (mass or molar) of a given species present in the mixture. The various chemical species (molecules) can be destroyed or created in a mixture owing to chemical reactions, however, the total number of individual types of atoms present in the system are not altered. This fundamental principle is referred to as conservation of mass in a chemically reacting system [2, 5].

Knowing the temperature, pressure, composition, and flow rates at the inlets and outlets of units in a chemical manufacturing plant can provide an adequate data structure for comprehensively assessing the process information for that unit. For each inlet stream, we assume the substance contains m_i species, whereas for each outlet stream, the substance contains m_o species. For a given unit process, assuming there will be i inlet streams, and o outlet streams, one needs to perform a total of

$$N_{measurements} = i * (3 + m_i - 1) + o * (3 + m_o - 1)$$

measurements for a comprehensive analysis of the unit process. Figure 7.1 schematically shows these connections of streams into and out of a single process unit. Note that the three variables counted in default for each stream are the temperature (T), pressure (P), and flow rate (Q), whereas we only need to know the $m_i - 1$ compositions directly for a stream containing m_i species.

Fig. 7.1 Connections and number of process measurements for a single unit process in chemical manufacturing

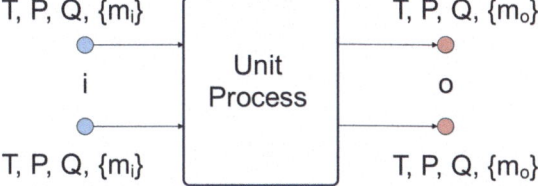

7.3.2 Transport Phenomena Modeling

Using fundamental principles of mass, momentum, and energy conservation, transport models are developed to describe the movement of species and energy within the process [3, 6]. These models involve:

- Mass balance equations for tracking species flow rates and concentrations.
- Momentum balance equations governing pressure drop calculations and fluid dynamics.
- Energy balance equations for quantifying heat transfer and energy consumption.

Incorporating appropriate boundary conditions and thermo-physical properties in these models ensures an accurate representation of transport phenomena in process components.

7.3.3 Chemical Reaction Kinetics and Thermochemistry

When chemical reactions occur, reaction kinetics and thermochemical principles must be considered. Rate laws, Arrhenius-type expressions, and reactor design equations help assess reaction rates and process dynamics. Additionally, thermochemical calculations involving enthalpy of reaction and equilibrium constants provide insights into reaction feasibility and energy requirements. By integrating these aspects, a holistic understanding of the process dynamics is achieved.

Modeling chemical kinetics in different environmental and industrial processes requires specialized methodologies and databases that captures reaction mechanisms, transport processes, and equilibrium calculations. For aqueous geochemistry, software like PHREEQC [7] is widely used to model chemical speciation, solubility equilibria, and reaction kinetics in natural and engineered water systems. These models use equilibrium-based calculations with thermodynamic databases to predict metal precipitation, adsorption, and carbonate equilibria, making them invaluable for water quality and remediation studies.

In combustion kinetics, tools like Cantera [8] enable the simulation of chemical reaction mechanisms, flame propagation, and pollutant formation in combustion systems. Cantera provides an object-oriented platform for solving coupled chemical kinetics, thermodynamics, and transport processes, making it a useful tool for designing energy-efficient combustion engines and alternative fuel research.

For biochemical reaction modeling in wastewater treatment, software such as BioWin [9] is used to simulate activated sludge processes, anaerobic digestion, and nutrient removal. These models incorporate Monod-based reaction kinetics, mass balances, and microbial interactions to optimize wastewater treatment plant operations and improve effluent quality. The integration of such kinetic models into process control strategies enhances predictive capabilities and decision-making in environmental engineering.

7.3 From Data to Process Information

As described earlier, streams have properties that depend on the temperature, pressure, flow rate, as well as their material composition. Of these, while tracking pressure, temperature, and flow rate of each stream is relatively straightforward, it is considerably involved to track the different types of chemical species that are formed or destroyed in each process (particularly in reacting systems). Taking the example of a combustion reactor, consider the combustion of methane (CH_4) with air (O_2 + 3.76 N_2) in a steady flow reactor. In this case, there will be two input streams to the reactor node, namely, methane and air. The combustion products, however, will contain a larger number of species, with carbon dioxide (CO_2) and water (H_2O) being the main combustion products, along with unreacted species (N_2). Depending on the conditions of combustion, additional species may form (NO, NO_2, CO, *etc.*). Therefore, the combustion product stream can be an array of chemical species that is substantially different from the species present in the input streams. This is an important feature of reacting systems where the reactor nodes will have transformation properties enabling creation of new types of species in the product streams. In contrast, non-reacting units or nodes, such as separation processes, will not result in creation of new species.

A generalized approach of treating chemical species formation and destruction for a specific unit process (node), particularly those with reaction, involves creating an initial species superset array that contains all the species reported in the streams entering and exiting the unit. The composition of species in each stream entering and exiting the unit can then be mapped as mass or mole fractions based on the constituents of this superset. Species absent in any stream are assigned mass or mole fractions of zero.

7.3.4 Physics-Based Digital Twin Modeling of Process Components

Developing physics-based digital twins for chemical process components relies on a fundamental understanding of thermodynamic properties, transport models, conservation equations, and chemical kinetics. These elements form the core framework for accurately simulating individual unit operations and predicting their operational behavior under various conditions. The digital twin of a process component integrates design specifications, first-principles models, and real-time plant data to assess performance, optimize efficiency, and diagnose deviations from expected operation.

A shell-and-tube heat exchanger serves as a classical example of a digital twin application. A detailed model of the exchanger incorporates enthalpy balances, conductive and convective heat transfer equations, and transport correlations to predict temperature profiles, pressure drops, and heat transfer effectiveness (Fig. 7.2). By using design parameters such as shell diameter, tube arrangement, and flow rates, along with actual plant data (inlet/outlet temperatures, flow rates, and pressure readings), the digital twin can detect fouling, scaling, or reduced heat transfer efficiency. The model also allows predictive maintenance by estimating when cleaning or retrofitting is necessary.

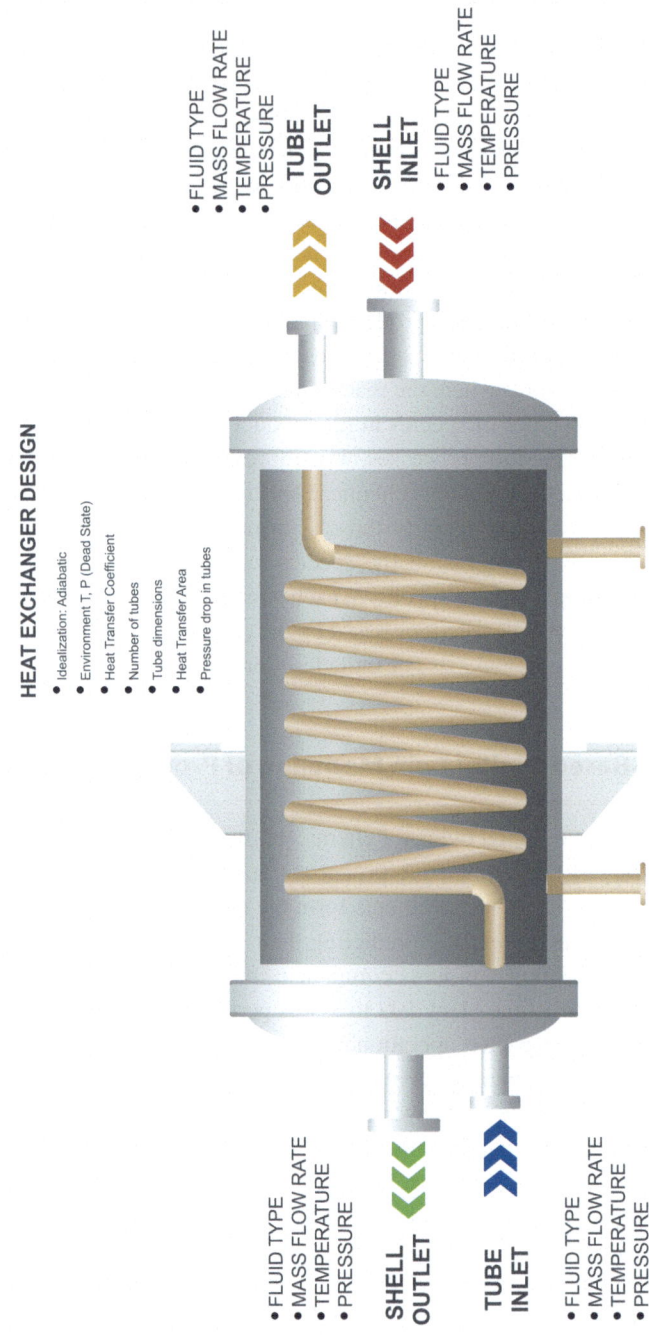

Fig. 7.2 Illustration of a shell-and-tube heat exchanger model used in a digital twin framework

Similarly, a natural gas-fired compression/spark-ignition engine with a turbocharger assembly benefits from a reaction-kinetic model based digital twin. Such a model incorporates combustion kinetics, thermodynamic state equations, and species transport equations to simulate air-fuel mixing, flame propagation, and exhaust gas composition. The reaction kinetics of methane combustion, for example, dictate the formation of CO_2, H_2O, NO_x, and unburned hydrocarbons under varying air-fuel equivalence ratios. By using measured cylinder pressures, temperatures, exhaust gas composition, and turbocharger speeds, the digital twin evaluates combustion efficiency, turbo lag, and potential emissions non-compliance.

By fusing physics-based models with real-time sensor data, digital twins offer deep process insight, anomaly detection, and predictive control capabilities. This approach enables condition monitoring, optimization, and fault diagnosis, ensuring that process components operate at peak efficiency while minimizing downtime and energy consumption.

7.3.5 Network Modeling of Process Components

A plant-wide model is constructed to capture the dynamic interactions between the unit operations and establishing the connectivity of material and energy flows between these units. This network representation integrates:

- Process units as nodes with individual dynamics.
- Flow paths of mass and energy as directed edges.
- Feedback loops and control interactions to assess process stability and performance.

The network modeling approach involves defining each unit process as a node, and the various streams as paths in a network graph. Each node will have a "transformative" role on the materials entering them, converting them to the output streams. For instance, a steam boiler is a specific type of transformer that will convert a compressed liquid to a compressed vapor by adding heat to the liquid. Reactors are another type of nodes with properties of reorganizing the molecular organization of atoms in the reactants entering them to form different types of chemical species in the product streams. for instance, a combustion reactor (a cylinder of an engine) combines a hydrocarbon fuel (natural gas) and air and transforms them to carbon dioxide and water vapor, thereby releasing a substantial amount of thermal energy, which can be converted to work. These transformative behaviors of individual unit processes can be modeled mathematically, with conservation of mass and energy (also momentum) forming the essential basis of performing such modeling.

This unified framework enables performance mapping against plant design specifications, helping identify operational inefficiencies and optimization opportunities. For example, Fig. 7.4 depicts how the spark ignition engine digital twin of Fig. 7.3 is embedded in a network of several units representing fuel gas mixing, air compression and preheating, exhaust management, and engine thermal management. These individual process digital

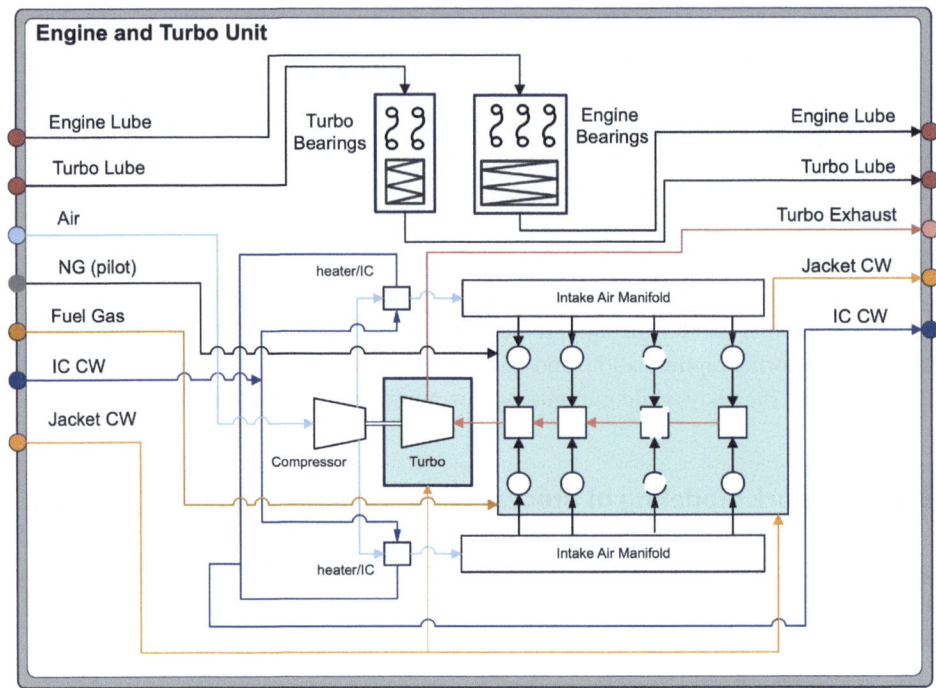

Fig. 7.3 Illustration of a digital twin abstract process model for a spark ignition engine with a turbo charger

twins are networked with the streams of different materials (and energy) connecting these. It is interesting to note that the overall system itself can be viewed as a lumped system (blackbox) with material inputs and outputs delineated appropriately. This demonstrates how a physics based modeling approach can be utilized with network models and graph theoretic modeling approaches to construct a network of multiple unit processes in a sequence to model a complete chemical manufacturing plant.

The fundamental premise of any chemical manufacturing is that raw materials enter as streams into unit processes where they are "processed" physically or chemically into product streams. The process shown in Fig. 7.1 or the overall bounding box of Fig. 7.4 could both represent an entire plant with raw materials as input streams, and the useful products and wastes as the output streams. In addition to the material flows, energy needs to be supplied and extracted from the plant. Energy also flows into and out of the plant through the material streams. Therefore, the construct of Fig. 7.1 is a simple block representation of a plant at any level, whether it is the macroscopic representation of the full plant, or the representation of a single processing unit of a plant. These blocks can be assembled into a network of blocks

7.3 From Data to Process Information

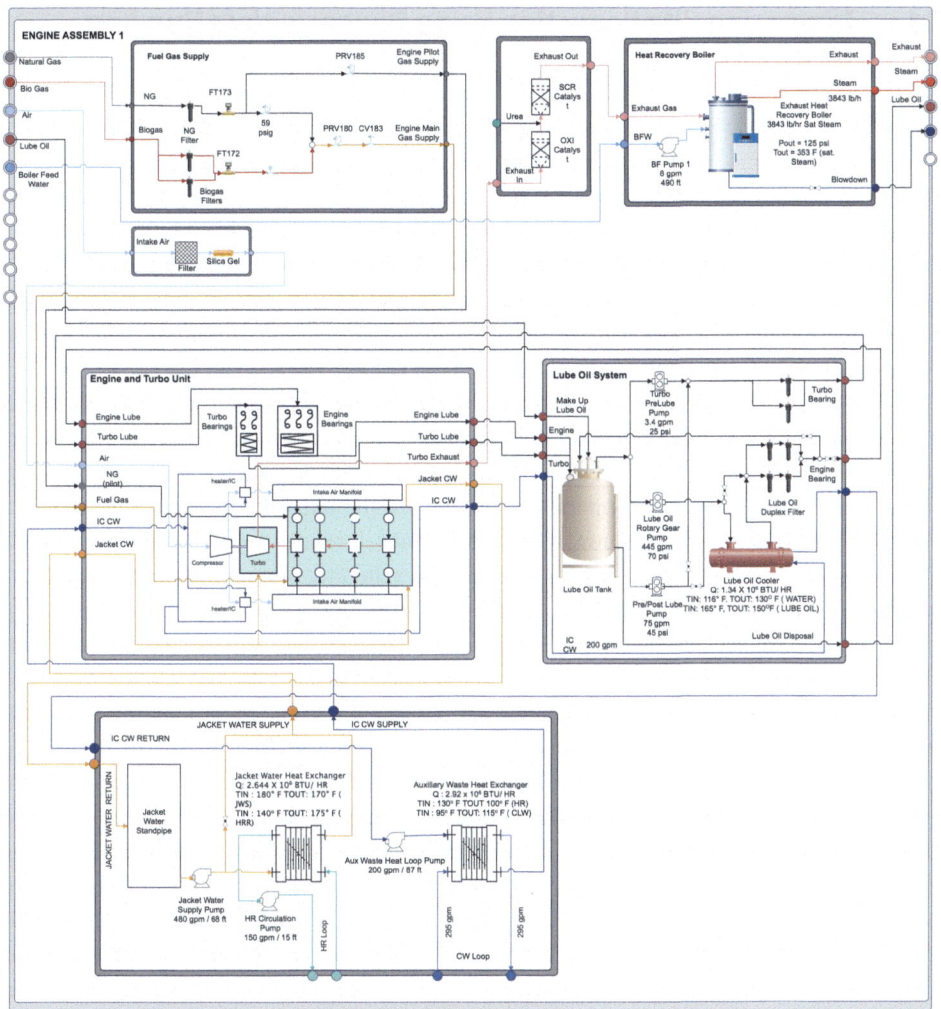

Fig. 7.4 Illustration of a network of digital twin process models for the spark ignition engine shown in Fig. 7.3 along with other digital twins for ancillary fuel gas supply, air supply, thermal management, and exhaust management components. The overall system itself can be considered as a single unit with material and energy inputs and outputs as in Fig. 7.1

and streams to build a complex plant involving multiple processing steps (Fig. 7.4). This type of generic structure forms the essential network structure of most chemical processing plants.

7.4 Dimensionality Reduction Using Dimensionless Numbers

Data-driven modeling approaches, such as unsupervised and supervised machine learning require large datasets to capture the complexity of chemical processes [10]. The number of process variables monitored in a chemical processing plant can be substantial, often comprising thousands of sensor tags. Statistical data-science methods employ algorithms such as classification, regularization, and regression to extract correlations from high-dimensional datasets [10]. However, these techniques demand large datasets, thereby amplifying the three Vs of big data: velocity, volume, and variety. The high data requirements of machine learning algorithms further exacerbate the computational burden for training and necessitate extensive digital infrastructure investments.

To mitigate these challenges, dimensionality reduction techniques are often employed. One widely used statistical approach is *Principal Component Analysis* (PCA) [11], which transforms the original high-dimensional dataset into a lower-dimensional space by identifying orthogonal principal components that capture the maximum variance. While PCA effectively reduces the data volume, it lacks physical relevance; the derived principal components are statistical constructs rather than variables directly interpretable in the context of process engineering. Despite this limitation, PCA and similar methods decrease data storage requirements, enhance computational efficiency, and reduce hardware costs associated with data acquisition and processing.

In contrast to purely statistical approaches, physics-based process modeling inherently provides natural dimensionality reduction through the use of *dimensionless groups* [12]. The fundamental equations governing chemical processes—such as mass, momentum, and energy conservation—naturally give rise to dimensionless numbers, which capture essential system dynamics while reducing the number of independent variables. Table 7.2 presents key dimensionless numbers in process engineering along with their significance. Classical examples include the Reynolds number (Re), which characterizes flow regimes, the Prandtl number (Pr), which relates momentum and thermal diffusivity, and the Damköhler number (Da), which compares reaction kinetics to transport rates. These groups offer meaningful physical insights and reduce the volume and velocity of data necessary for process analysis. A conventional yet underutilized approach in data-driven process analytics is the application of dimensionless groups. These numbers offer valuable insights into scaling, similarity, and dominant physical effects in chemical processes. By employing dimensionless analysis, large datasets can be reduced to essential parameters, facilitating model simplification and improved interpretation of process behavior.

By integrating both data-driven and physics-based approaches, a more structured methodology for plant data analysis emerges. For instance, while machine learning models can be trained on extensive datasets to identify complex patterns, embedding fundamental dimensionless groups as input features constrains the model within physically meaningful bounds, improving interpretability and reducing unnecessary data complexity. A balanced approach thus enhances predictive accuracy, reduces computational overhead, and facilitates the gen-

Table 7.2 Common dimensionless groups in process engineering

Name	Symbol	Equation	Description
Reynolds Number	Re	$\dfrac{\rho u L}{\mu}$	Ratio of inertial to viscous forces in fluid flow
Prandtl Number	Pr	$\dfrac{c_p \mu}{k}$	Ratio of momentum diffusivity to thermal diffusivity
Schmidt Number	Sc	$\dfrac{\mu}{\rho D}$	Ratio of momentum diffusivity to mass diffusivity
Péclet Number	Pe	$Re \cdot Pr$	Measures relative importance of advection to diffusion
Nusselt Number	Nu	$\dfrac{hL}{k}$	Ratio of convective to conductive heat transfer
Sherwood Number	Sh	$\dfrac{k_m L}{D}$	Analogous to Nusselt number for mass transfer
Damköhler Number	Da	$\dfrac{kL}{u}$	Compares reaction rate to transport rate
Biot Number	Bi	$\dfrac{hL}{k}$	Ratio of internal to surface thermal resistance
Fourier Number	Fo	$\dfrac{\alpha t}{L^2}$	Governs transient heat conduction

eration of correlations with lower data volume and velocity requirements. This synergy between data-driven analytics and first-principles modeling presents an efficient and rational strategy for digital twin development and process optimization in the chemical industry.

7.5 Hybrid Data- and Physics-Driven Approaches

This chapter outlined a framework for chemical process data analysis, emphasizing thermodynamic and transport properties, reaction kinetics, and network-based process modeling. The use of dimensionless groups in process analytics provides a promising research direction, enabling effective data reduction and enhancing the development of digital twins. Future work should explore advanced techniques integrating data-driven and physics-based models for improved process optimization and control.

The design, construction, and operation of chemical process plants are deeply rooted in the principles of chemical engineering and process science. Over decades of industrial evolution, the collective technical knowledge of reaction kinetics, transport phenomena, thermodynamics, and control engineering has been meticulously refined and integrated into such a plant's design. This embedded intelligence crafted through rigorous analysis, empirical validation, and engineering best practices constitutes the fundamental DNA of the plant's operation. To overlook this intrinsic foundation in favor of a purely data-driven digitalization approach would not only be impractical but could also result in significant inefficiencies and an arduous, redundant learning process.

A purely data-driven approach, wherein digital models attempt to infer process behavior solely from historical plant data, often requires extensive time and massive datasets to establish reliable correlations. These methods, though powerful in pattern recognition, lack inherent physical intuition and may fail in extrapolating to conditions outside the trained dataset. Moreover, the complexity of chemical plants where interactions between variables are governed by well-established physical and chemical laws renders an exclusive reliance on machine learning models suboptimal, especially in cases requiring technical interpretation of the results and rationalization of predictions. In the absence of the embedded technical knowledge, there is no other means of processing a plant data except for utilizing statistical data-processing algorithms and machine learning to create knowledge and wisdom. However, it is overambitious to expect artificial intelligence (AI) can "figure out the underlying process dynamics from the data".

In contrast, digital twin models offer a more structured and effective pathway toward digital transformation. A digital twin is simply a mathematical model of the process that captures its essential operational features and performance dynamics. By integrating fundamental process knowledge with real-time plant data, digital twins provide a physics-informed framework for data aggregation, analysis, and decision-making. Incorporating first-principles models, such as mass and energy balances, reaction kinetics, and transport equations, ensures that data-driven insights remain consistent with known physical behaviors. Furthermore, leveraging the process dynamics inherent in the plant's design allows for a more rational approach to digitalization. Instead of blindly rediscovering operational patterns, digital twin models enable operators and engineers to contextualize anomalies, predict failures, and optimize performance with a fraction of the effort required by purely statistical methods.

While physics-based digital twin models provide a structured foundation for process analysis and optimization, they are inherently limited by the approximations and simplifications in fundamental models. All first-principles approaches rely on assumptions whether in thermodynamic equations of state, reaction kinetics, or transport models, that, while effective, may not fully capture the intricate dynamics of real-world process systems. This is where machine learning (ML) and artificial intelligence (AI) serve as powerful complementary tools. By analyzing vast amounts of plant data, ML algorithms can uncover additional features, detect hidden correlations, and reveal subtle process interactions that may not be explicitly modeled in first-principles equations. These data-driven insights can be invaluable in identifying plant-specific behaviors, diagnosing inefficiencies, and even discovering new optimal operating conditions beyond conventional process constraints. Moreover, AI-driven optimization techniques can refine process settings by continuously learning from plant operations, adapting to variability, and improving efficiency over time.

Hybridizing physics-based models with ML techniques offers the best of both worlds—ensuring process integrity through fundamental laws while leveraging data-driven adaptability to capture plant-specific nuances. This synergy allows for a more intelligent and adaptive digital twin, one that not only predicts performance based on known physical principles but also evolves through continuous learning. This hybrid approach not only enhances predictive accuracy but also significantly reduces the amount of data required to develop reliable models, thus lowering computational costs and simplifying implementation. By integrating ML and AI within digital twin frameworks, process industries can move beyond static modeling and towards a more dynamic, self-optimizing digitalization approach, achieving greater resilience, efficiency, and insight into complex industrial systems.

While digitalization is being implemented in many organizations and sectors of business, a key question becomes, what internal capacity building, retraining, and changes will be needed in these organizations to enable them to handle and properly capitalize on the benefits of the digital transformation.

References

1. A.V. Oppenheim and R.W. Schafer. *Digital Signal Processing*. Prentice Hall international editions. Prentice-Hall, 1975.
2. Herbert B. Callen. *Thermodynamics and an Introduction to Thermostatistics*. John Wiley & Sons, 2nd edition, 1985.
3. R. Byron Bird, Warren E. Stewart, and Edwin N. Lightfoot. *Transport Phenomena*. John Wiley & Sons, 2nd edition, 2007.
4. J. M. Smith, H. C. Van Ness, M. M. Abbott, and M. T. Swihart. *Introduction to Chemical Engineering Thermodynamics*. McGraw-Hill Education, 8th edition, 2017.
5. Yunus A. Çengel and Michael A. Boles. *Thermodynamics: An Engineering Approach*. McGraw-Hill Education, 9th edition, 2019.
6. Jacob Masliyah and Subir Bhattacharjee. *Electrokinetic and Colloid Transport Phenomena*. John Wiley & Sons, 2006.
7. David L. Parkhurst and C. A. J. Appelo. *Description of Input and Examples for PHREEQC Version 3 - A Computer Program for Speciation, Batch-Reaction, One-Dimensional Transport, and Inverse Geochemical Calculations*. U.S. Geological Survey, 2013.
8. David G. Goodwin, Harry K. Moffat, and Raymond L. Speth. *Cantera: An Object-Oriented Software Toolkit for Chemical Kinetics, Thermodynamics, and Transport Processes*. Zenodo, 2016.
9. Envirosim Associates Ltd. *BioWin Technical Reference*. Envirosim Associates, 2003.
10. A. Ananthaswamy. *Why Machines Learn: The Elegant Maths Behind Modern AI*. Penguin Books Limited, 2024.
11. I. T. Jolliffe. *Principal Component Analysis*. Springer Series in Statistics. Springer, New York, NY, 2 edition, 2002.
12. Jonathan Worstell. *Dimensional Analysis: Practical Guides in Chemical Engineering*. Butterworth-Heinemann, 2014.

Part III
Corporate Vision and Human Aspects of Digital Transformation

8 Digital Transformation–Corporate Vision

8.1 Toward Industrie 5.0

The journey of industrial transformation has brought us from mechanization in first Industrial Revolution to the digital and interconnected world of Industrie 4.0. Each of these stages has aimed to push the boundaries of productivity, efficiency, and connectivity. However, as we navigate the fourth industrial revolution, an emerging paradigm is on the horizon: Industrie 5.0. Unlike its predecessors, Industrie 5.0 is poised to shift its focus from purely technology-driven outcomes to a model that embraces human-centricity, sustainability, and resilience [1, 2]. For the chemical processing industry, this transition introduces both challenges and transformative opportunities that promise a more balanced and purpose-driven manufacturing approach.

While Industrie 4.0 has focused on advancing automation, data analytics, artificial intelligence, and IoT to optimize production systems, Industrie 5.0 centers on placing humans and sustainability at the heart of technological transformation. By doing so, this new paradigm creates a manufacturing ecosystem where technological advancements support, rather than replace, human ingenuity, and where processes operate in harmony with environmental and societal considerations [3]. This chapter explores the core principles of Industrie 5.0 in the context of chemical processing, focusing on how human-centricity, sustainability, and resilience can reshape the future of the industry.

8.1.1 Human-Centricity: Redefining the Role of Technology for the Workforce

The first and most prominent principle of Industrie 5.0 is human-centricity. In the industrial context, human-centricity emphasizes designing technology and systems that cater to the needs, skills, and growth of the human workforce. For years, the narrative surrounding digital transformation has focused on how workers can adapt to technology—learning new skills, optimizing machine performance, or managing complex digital interfaces. In contrast, Industrie 5.0 flips this approach, asking instead: "How can technology serve and empower the workers [4]?"

In the chemical processing industry, human-centricity has particular relevance as the work involves complex processes, potentially hazardous materials, and intricate operations that require precise oversight. Embracing a human-centric approach means building digital solutions that not only enhance productivity but also improve job satisfaction, worker engagement, and overall well-being.

8.1.2 Moving from Task Automation to Decision Support

With Industrie 5.0, the aim is not merely to automate repetitive tasks but to enable workers to make informed decisions and focus on higher-level problem-solving. For instance, digital platforms powered by AI and machine learning can assist workers by analyzing real-time data and predicting equipment performance. These systems provide critical insights, allowing operators to make strategic decisions that improve process outcomes rather than simply monitoring outputs. In this scenario, workers evolve from task executors to decision-makers, empowered by digital tools that support, rather than replace, their expertise.

8.1.3 Redefining Job Roles and Skills

By centering technology around human capabilities, Industrie 5.0 opens the door for redefining job roles within chemical processing. Digital tools are designed not just for productivity gains but to enable new roles focused on innovation, system integration, and sustainable operations. For example, a plant operator's role may evolve from overseeing production to becoming an "operational strategist," responsible for optimizing processes using data-driven insights. This shift encourages workers to develop skills in data analysis, decision-making, and sustainable practices, ultimately leading to more fulfilling work.

In chemical processing, many manual roles still involve repetitive tasks, which can diminish job satisfaction and engagement. A human-centric approach to digital transformation addresses these concerns by elevating these roles to focus on strategic activities. Workers could engage in tasks that require critical thinking and creativity, such as troubleshooting issues, suggesting process improvements, or experimenting with eco-friendly materials and methods. By fostering this kind of engagement, human-centric design supports a culture of innovation and continuous improvement in the workplace.

8.1.4 Enhancing Worker Safety and Well-Being

Industrie 5.0 also brings a renewed emphasis on worker safety and well-being. In chemical processing, where operators are often exposed to harsh environments, hazardous materials, and physically demanding tasks, digital solutions can significantly enhance safety standards. Wearable technology, for instance, can monitor vital signs and environmental conditions, alerting workers and supervisors to potential health risks. Additionally, augmented reality (AR) systems could assist in providing real-time guidance, reducing the cognitive load during complex tasks and minimizing the chances of error.

Overall, the human-centric focus of Industrie 5.0 seeks to create a work environment where technology enables, rather than controls, the workforce. This shift not only enhances productivity but also nurtures a culture of mutual respect between humans and technology, ultimately contributing to a more engaged, capable, and satisfied workforce [4].

8.2 Sustainability: Expanding the Scope of Industrial Responsibility

As industries have advanced through various stages of industrialization, sustainability has often been viewed as a supplementary goal rather than a core principle. Industrie 5.0 challenges this paradigm by embedding sustainability into the very fabric of manufacturing processes [3]. In the chemical processing industry, sustainability takes on a dual meaning: minimizing environmental impacts and optimizing resource use while also considering the broader implications of industrial activities on society and ecosystems.

8.2.1 Adopting a Holistic Approach to Resource Management

Industrie 4.0 has brought about significant advancements in energy and material efficiency, largely through optimization of discrete processes or individual units. Industrie 5.0, however, encourages an integrated approach that considers the entire plant's resource usage holistically. This approach acknowledges that the various units of a chemical processing facility—whether water management, energy consumption, or waste treatment—are interconnected, and that optimizing one aspect can often lead to inefficiencies elsewhere.

For example, water reuse and recycling are already prevalent in many chemical plants, but Industrie 5.0 expands the scope of water management by evaluating its integration with other resource systems. A chemical plant can benefit from an interconnected view where water, energy, and raw materials are managed as a whole. Such a system can leverage digital tools like data analytics to identify patterns, predict resource demands, and suggest adaptive measures that optimize resource use across the facility rather than focusing solely on individual units.

8.2.2 Embracing Circularity Beyond the Plant Boundary

While many current sustainability efforts are contained within the physical and operational boundaries of manufacturing plants, Industrie 5.0 advocates for an expanded view of circularity [3]. This expanded perspective encourages manufacturers to partner with external entities, such as municipalities, regional water authorities, or other industries, to create a circular economy that extends beyond a single plant. For example, certain municipalities are increasingly investing in water recycling infrastructure for indirect potable reuse, creating opportunities for industries to partner with these facilities rather than relying solely on their own systems.

Such partnerships can help chemical plants align their sustainability goals with broader community objectives, reduce costs, and gain access to shared resources. Additionally, by sharing digital insights on resource quality and usage patterns, manufacturers can make more informed decisions about resource procurement and disposal, ultimately fostering a collaborative approach to sustainability that benefits both the industry and the surrounding community.

8.2.3 Leveraging Data for Predictive Sustainability

Another key aspect of Industrie 5.0's approach to sustainability is its emphasis on predictive capabilities through data. In a chemical processing environment, where resource use and emissions can fluctuate with operational conditions, predictive analytics enable plants to anticipate and adapt to changes in real-time. AI-driven models can, for example, forecast water or energy usage patterns, allowing operators to make proactive adjustments that minimize waste and environmental impact. By focusing on prediction rather than reaction, Industrie 5.0 transforms sustainability from a compliance-based model to a proactive strategy that anticipates future needs and constraints.

8.3 Resilience: Building Robust Systems for an Uncertain Future

The concept of resilience, creating systems that can withstand and adapt to change, is critical to Industrie 5.0. Resilience involves designing systems that not only optimize for efficiency and productivity under normal conditions but also maintain stability during disruptions [3]. For the chemical processing industry, resilience is especially crucial given the volatility of raw material supply chains, stringent environmental regulations, and the risks associated with operating in uncertain global markets.

8.3.1 Redesigning Systems with Flexibility and Adaptability

In traditional industrial models, manufacturing systems are often optimized for maximum efficiency in stable environments, which can lead to vulnerabilities when disruptions occur. Industrie 5.0, however, encourages the design of flexible systems capable of adjusting to fluctuating conditions, such as changes in raw material availability or variations in demand. By incorporating adaptive algorithms and modular infrastructure, chemical plants can transition from rigid, high-risk systems to more agile operations that can reconfigure processes, optimize resources, and maintain continuity even in turbulent conditions.

8.3.2 Enhancing Supply Chain Resilience Through Digital Connectivity

Supply chain disruptions can have significant ripple effects on production, especially in the chemical processing industry, where timely access to specific raw materials is critical. Industrie 5.0 leverages digital connectivity and data transparency across supply chains to enhance resilience. For example, IoT-enabled tracking and blockchain technology can provide real-time visibility into raw material origins, transit status, and quality conditions, allowing plants to monitor and anticipate disruptions in advance. By establishing digital links throughout the supply chain, chemical processing facilities can respond quickly to potential bottlenecks, whether by adjusting production plans, sourcing from alternative suppliers, or reallocating resources.

8.3.3 Fostering a Resilient Workforce Through Cross-Training and Knowledge Sharing

In a human-centric Industrie 5.0 paradigm, workforce resilience is as important as operational resilience. By investing in cross-training and knowledge-sharing initiatives, chemical processing companies can cultivate a workforce that is adaptable and prepared to handle diverse roles and responsibilities. When workers have the skills to manage multiple functions or step into critical roles during emergencies, the plant as a whole becomes more resilient. Furthermore, knowledge-sharing platforms, augmented reality training, and virtual collaborative tools can create a culture of continuous learning that empowers employees to adapt

8.4 Digitalization: More than CapEX Approach

In the chemical processing industry, the scale of capital investment required for physical infrastructure—such as heavy machinery, reactors, and other specialized equipment—is substantial. This industry demands a unique set of physical assets tailored to specific processes and strict regulatory compliance, making these plants resource-intensive to set up and operate. These investments mean that any major process change is rare, involving long cycles of planning, approval, and deployment, and sometimes requiring years of preparation. For these reasons, digital transformation initiatives in such facilities are often approached through a capital expenditure (CapEx) lens. Digital transformation, in these cases, can be perceived as another large-scale project, with a focused team mobilizing toward its implementation.

However, while the CapEx approach is tried and true for physical assets, it is not necessarily well-suited to digital transformation, which operates on fundamentally different principles. Unlike heavy equipment, digital technologies—such as data analytics platforms, machine learning models, or IoT-enabled sensors—require rapid, iterative development and frequent updating. The hardware and software frameworks that power digital transformation are far more dynamic and adaptable than static, physical equipment. If digital transformation initiatives are treated as static capital projects, they can quickly become outdated, limiting their usefulness and hindering the plant's technological advancement.

Digital infrastructure is inherently different in that it evolves continuously. Unlike conventional machinery that may function optimally for decades with minimal upgrades, software and digital systems need regular updates to maintain effectiveness, security, and relevance. This constant evolution is a double-edged sword: while digital solutions can offer incremental improvements, they also introduce the necessity of frequent maintenance, upgrades, and often re-engineering as the technology landscape advances. Implementing a digital solution in a manufacturing plant should therefore not be treated as a one-time CapEx project but as an ongoing investment, more akin to operational expenditure (OpEx) that involves continual optimization. Embracing this OpEx mentality enables chemical plants to adapt their digital infrastructure seamlessly and stay abreast of technological advancements.

This OpEx approach also aligns with a mindset shift necessary for digital transformation success. Rather than focusing solely on initial deployment, success metrics should emphasize continuous improvement and adaptability. Manufacturing plants have long relied on a culture of continuous improvement for traditional processes—adjusting, calibrating, and refining procedures for optimal output. However, to succeed in digital transformation, plants must extend this mindset to technology. Continuous improvement in digital terms means iterating software, integrating new algorithms, refining data models, and possibly even adopting new digital solutions when better options arise.

A critical challenge in this shift is that digital transformation requires a different skill set than traditional manufacturing. Operations teams, accustomed to maintaining and troubleshooting heavy equipment, now also need to manage digital systems, which can be vastly different in terms of complexity, flexibility, and upkeep requirements. While physical machines have established maintenance cycles and predictable wear and tear, digital tools can fail in unexpected ways, or they may simply become obsolete. This necessitates both digital literacy and an adaptive approach in plant teams. As such, digital transformation success depends not only on the technology itself but also on the workforce's ability to adapt to, manage, and continuously optimize these technologies.

8.4 Digitalization: More than CapEX Approach

This difference between digital and traditional CapEx projects also has implications for how projects are planned, managed, and evaluated. Traditional capital projects follow a lifecycle of design, construction, commissioning, and handover to operations. However, in the digital space, this linear approach is replaced by a cyclic process. Digital solutions require iterative phases of development, testing, feedback, and refinement. Once a digital system is deployed, it needs continuous performance evaluation, iterative updates, and proactive troubleshooting. This cycle can initially be disconcerting for plant teams accustomed to the clear finish line of a traditional project, but it becomes essential for maintaining the relevance and effectiveness of digital tools.

Digital transformation also changes how success is measured. In CapEx projects, success is often judged based on adherence to budget, schedule, and scope, as well as the smooth handover of assets to operations teams. For digital projects, however, success is best understood through adaptability, seamless system integration, and sustained user satisfaction over time. This means that performance metrics should be established not only for the initial deployment but also for long-term engagement, effectiveness, and user experience. Without these considerations, even the best-designed digital solution may falter if users find it difficult to operate or if it lacks the flexibility to adapt to evolving operational needs.

Moreover, the CapEx approach typically encourages the formation of temporary project teams responsible for design and installation, who then hand over the project to operations once complete. For digital initiatives, this handover model is impractical and often counterproductive. Digital transformation thrives when project teams remain engaged long-term, enabling ongoing support, training, and troubleshooting. Operations staff benefit from this sustained collaboration, as they gain confidence and experience in managing digital systems. A continuous support model helps ensure the digital solution remains viable, valuable, and aligned with the plant's operational goals.

The frequent pace of change in digital technology means that the implementation strategy must also be flexible. Hardware and software are constantly advancing, with new updates, versions, and entirely new systems emerging in quick succession. The faster these technologies change, the greater the risk that a plant's digital infrastructure becomes outdated if it is treated as a fixed CapEx investment. Instead, by budgeting for periodic upgrades and maintaining a flexible digital roadmap, plants can ensure that their systems remain current. Regular reviews of digital infrastructure can identify when updates are necessary, whether for compatibility, security, or performance enhancements. In a rapidly changing technological landscape, this flexibility is not just an advantage—it is a necessity.

Additionally, the success of digital transformation depends on the integration of digital solutions with existing plant operations. Digital technologies are often implemented in silos, isolated from the rest of the plant's systems. While initial adoption may be simpler in a siloed model, it can limit the digital solution's impact, leading to missed opportunities for comprehensive data analysis and coordinated decision-making. Integrating digital tools with traditional systems creates a seamless flow of information across the plant, enabling more holistic operational insights. However, this integration requires careful planning, as well as a willingness to modify legacy systems and adapt processes to accommodate new digital capabilities.

The rapid pace of digital innovation also presents a risk of over-investing in technology that may become obsolete within a short period. For instance, adopting a specific data analytics platform may seem like a good investment, but if a more advanced solution emerges within a year, the initial investment could become a sunk cost. By adopting a modular approach to digital transformation, plants can mitigate this risk. Modular implementation involves selecting flexible, scalable systems that allow for incremental upgrades and modifications without disrupting the entire digital infrastructure. This modularity also supports interoperability, ensuring that new technologies can integrate with existing systems as they are introduced.

A shift in mindset is essential for both plant leaders and employees to view digital transformation as a dynamic journey rather than a one-time upgrade. The nature of digital technologies, with their frequent updates and rapid advancements, does not align with the static nature of traditional CapEx projects. Viewing digital transformation through an OpEx lens encourages ongoing investment in training, maintenance, and technology upgrades, all of which are crucial for sustained success. Leaders play a pivotal role in this shift, as they must advocate for a culture that embraces change, continuous learning, and adaptation. This cultural shift, while challenging, can be a catalyst for achieving higher productivity, operational efficiency, and competitive advantage.

8.5 Digital Transformation: Implementing the Human Element

Digital transformation in chemical manufacturing is not just about adopting new technologies; it represents a fundamental change in how people within the organization work and collaborate. From the boardroom to the plant floor, the successful adoption of digital tools depends on an active and engaged workforce. The human element is essential for turning new technologies into meaningful improvements in efficiency, safety, and productivity.

8.5.1 The Corporate Goal and the Importance of Communication

Most digital transformation projects are guided by a corporate goal, often aiming to enhance efficiency, sustainability, and competitiveness. For this vision to be realized, leaders must clearly communicate these objectives across the organization to ensure alignment and secure employee buy-in. However, communicating the reasons behind the transformation can be more challenging than expected, especially in the chemical manufacturing industry, where long-established processes and resistance to change can create significant obstacles.

The key to overcoming these challenges lies in presenting the transformation in a way that resonates with employees at all levels. For example, emphasizing how digital tools can improve safety protocols or simplify complex tasks makes the benefits more tangible for workers who might otherwise view the transformation as unnecessary or even as a threat to their jobs.

This is especially crucial for frontline employees, who will be the primary users of these digital systems. When they understand how the new tools will improve their daily tasks, the chances of successful adoption increase. Creating a clear connection between the overall corporate vision and the personal benefits for employees is vital to fostering a positive attitude toward change.

8.5.2 Three Key Personas in Digital Transformation

In a chemical manufacturing setting, three major personas are critical to the success of digital transformation:

Strategy Setters: These are the leaders who define the corporate goals and shape the vision for digital transformation. They ensure the strategy aligns with the company's objectives, such as sustainability, efficiency, and competitiveness. Clear, transparent communication from strategy setters is crucial to convey the purpose of the transformation and how it will benefit the organization and its people.

Implementation Teams: This group includes IT specialists, engineers, and project managers responsible for turning the corporate strategy into action. They manage the integration of new digital tools, ensuring that cross-functional collaboration happens smoothly. Implementers act as a bridge between leadership and frontline employees, translating high-level goals into practical solutions and addressing any technical challenges along the way.

Frontline Employees: The frontline workers who use the digital tools daily are the backbone of the transformation. Their involvement is key to success, as they bring firsthand insights into how new systems will affect their day-to-day work. Engaging them early, incorporating their feedback, and providing proper training ensure that digital systems meet their needs and are fully adopted into regular operations.

By aligning these personas—strategy setters, implementers, and frontline employees—organizations can ensure that digital transformation is not only technologically successful but also embraced by the people who will drive it forward.

8.6 Strategy Setters: Crafting an Actionable Plan for Digital Transformation

8.6.1 Aligning Transformation with Corporate Strategy

A successful digital transformation hinges on a clear alignment between the transformation strategy and the broader corporate strategy. This alignment is not just beneficial; it's essential. Every facet of the transformation strategy must support the corporate strategy's key performance indicators (KPIs) and long-term objectives. Without this synchronization, the transformation risks deviating, causing inefficiencies or missed targets. It is crucial that the transformation strategy not only meets these KPIs but also clarifies the timeline and likelihood of success, addressing the inherent urgency of corporate goals.

By linking transformation efforts directly to corporate priorities, strategy setters ensure that initiatives avoid silos and meaningfully contribute to the organization's overall vision. This alignment also enables the effective allocation of resources, ensuring that digital investments are targeted where they have the greatest impact. Thus, before honing in on individual projects or site-specific solutions, the strategy setters must establish a strong connection between transformation goals and corporate ambitions.

8.6.2 Crafting a Comprehensive and Scalable Strategy for Implementation

Once alignment is secured, the challenge for strategy setters is to craft an actionable roadmap that outlines every phase of the digital transformation. This roadmap must address financial constraints, justify return on investment (ROI), and provide a realistic timeline for execution. A crucial element of this plan is scalability. While many digital transformation efforts begin with a pilot phase, the true test is whether the initiative can be scaled effectively across different sites, departments, or regions. Strategy setters must ensure that solutions designed for a smaller scale can expand without significant overhauls as the project grows.

Securing funding at the outset is critical to the success of this process. Underfunded projects may struggle to scale or adapt to new challenges. In digital transformation, scaling is more than just replicating solutions across sites—it requires thoughtful planning to maintain performance, effectiveness, and integration as the project expands. Without scalability, even the most successful pilot projects can falter when rolled out on a larger scale.

Strategy setters play a crucial role in crafting a roadmap for implementation that takes into account financial implications, justifies ROI, and outlines a clear timeline for execution. This roadmap must cover all aspects of the transformation, from vendor selection and technology deployment to budget allocation and employee training. The plan must be comprehensive yet flexible enough to adapt to the organization's evolving needs.

One significant challenge is ensuring that the strategy is actionable. High-level plans with broad goals are not enough. The strategy must include specific steps and milestones that can be implemented at each manufacturing site. This requires a deep understanding of the unique challenges and opportunities at each location, as well as the ability to tailor solutions to meet the specific needs of each site.

8.6.3 Balancing Business and Technical Considerations

Another critical role of strategy setters is to balance the business objectives of the organization with the technical requirements of digital transformation. While senior leaders often excel in business strategy, digital transformation requires a deep understanding of existing infrastructure, technology landscapes, and the cross-functional nature of the work involved.

A cross-functional approach is essential, bringing together operations, IT, finance, and other departments to create a strategy that aligns with both business and technical needs. This collaboration ensures the transformation is both feasible and scalable across the organization.

8.6.4 Vendor Selection and Technological Advancements

Vendor selection is another critical factor. Strategy setters are responsible for evaluating potential vendors based on their technical capabilities, expertise, and ability to meet the organization's specific needs. This process is closely tied to the rapid pace of technological advancement. As new digital tools and solutions emerge, strategy setters must continuously scout for advancements to ensure that chosen solutions remain relevant and competitive.

In addition to selecting the right vendors, strategy setters must also assess risks, from technical failures and cybersecurity threats to budget overruns and employee resistance. By identifying and mitigating these risks early, the strategy setters can increase the likelihood of a successful digital transformation.

8.6.5 The Role of Sponsors and Senior Decision-Makers

In any transformation that spans multiple functions and departments, differing priorities and opinions are inevitable. While healthy debate can improve decision-making, prolonged disagreements can stall progress. That's why having a senior sponsor or decision-maker assigned to the project is critical. This individual acts as a champion for the transformation, ensuring alignment and making decisive calls when necessary.

The sponsor also provides the project with legitimacy and authority, overcoming resistance and helping to secure the necessary resources. They serve as advocates within the organization, ensuring that the transformation remains aligned with corporate objectives. Sponsors play a pivotal role in navigating roadblocks and keeping the project on track, especially when conflicting interests arise between departments.

8.6.6 The Role of Communication in Sustaining Momentum

Digital transformation projects can be complex and slow to show tangible results. Effective communication becomes essential in sustaining momentum, as it reassures stakeholders that progress is being made. Transparency is key to maintaining confidence in the initiative, and one of the most effective ways to achieve this is through regular updates using dashboards, scorecards, and visual tools that translate complex data into clear, actionable insights.

These tools provide visibility into milestones, demonstrating progress even when larger outcomes remain on the horizon. Regular communication, tied to KPIs and strategic goals, ensures that leadership and employees alike stay engaged and motivated throughout the transformation process. Effective communication also fosters engagement across all levels of the organization, from frontline employees to senior leadership, helping everyone understand how their efforts contribute to the broader initiative.

8.6.7 Ensuring Continuity in Strategy Teams

A common challenge in large-scale digital transformations is the loss of continuity within the strategy-setting teams. Often, once the initial plan is drafted, the team responsible moves on, leaving execution to others. This handoff can result in miscommunication, resistance to adjustments, and a lack of adaptability. To avoid this, it is vital to treat the strategy as a living document that evolves as the project progresses.

Continuity allows the original team to stay involved throughout the implementation, maintaining the integrity of the initial vision while also allowing for adjustments based on real-world feedback. By keeping key team members engaged, the project can benefit from their deep understanding of the corporate vision and the intricacies of the transformation. This not only enhances problem-solving but also ensures the strategy remains adaptable to changes in technology, market conditions, or organizational needs.

8.7 Project Implementers: Building Cross-Functional Collaboration

8.7.1 Clear Corporate Goals to Successful Implementation

Once the corporate goal is established, the next step is implementation, where the human element becomes particularly significant. The implementation team, often made up of IT specialists, engineers, plant managers, and other functional experts, plays a vital role in translating the digital strategy into reality. However, for the project to be executed effectively, cross-functional collaboration is crucial.

8.7.2 IT-OT Integration: A Cornerstone of Digital Transformation

In the chemical manufacturing industry, digital transformation requires seamless integration between operational technology (OT) and information technology (IT). Traditionally, OT teams manage physical systems like machines, sensors, and production lines, while IT teams handle the company's digital infrastructure, including networks and data systems. The convergence of these two domains, referred to as IT-OT integration, is essential for any digital initiative to succeed.

For example, predictive maintenance tools rely on IT systems to capture and process data from OT systems in real-time, allowing the detection of equipment issues before they cause significant disruptions. Without proper collaboration between these teams, the project may suffer from mismatched expectations or poor system integration, ultimately leading to technical hurdles or project failure.

8.7.3 Cross-Functional Collaboration: Beyond IT and OT

Effective digital transformation involves more than just IT and OT collaboration. Other critical cross-functional teams, such as quality, environmental health and safety (EHS), procurement, and transportation, play equally significant roles. Each of these groups has different expectations and requirements that must be addressed within the digital transformation strategy.

For example, the quality team depends on reliable, precise digital tools to maintain product standards, while procurement may prioritize cost efficiency and supplier integration. If these teams' feedback isn't integrated, or if the digital tools fail to meet their needs, it can negatively affect their productivity, creating inefficiencies or even new risks. Therefore, a holistic approach to digital transformation that incorporates the needs and input of all functional teams is essential for success.

8.7.4 Agile Project Management

Agile project management practices are well-suited for digital transformation initiatives, emphasizing iteration, flexibility, and collaboration. Agile allows teams to work in short sprints, test new solutions on a smaller scale, and adjust quickly based on real-time feedback. This method is especially beneficial for complex environments like chemical manufacturing, where errors in system integration can lead to costly production delays or safety hazards.

However, chemical manufacturing sites often operate with rigid project management structures. Introducing agile methodologies helps these teams embrace a more iterative approach, improving responsiveness and speeding up project timelines. Agile methods enable teams to address challenges early in the process, ensuring solutions are both effective and safe before full-scale deployment.

8.7.5 Managing Legacy Systems and Technical Debt

One of the key challenges in digital transformation is managing legacy systems that may be outdated or incompatible with new technologies. Many chemical manufacturing plants carry a significant amount of technical debt—old systems that need to be updated or replaced before any new digital initiatives can be implemented.

Addressing this technical debt must be carefully planned, as substantial costs can arise from bringing outdated systems up to par before adding any new technologies. Failure to tackle this issue can lead to project delays, budget overruns, or even failure. Therefore, it is crucial to weigh the long-term benefits of upgrading existing systems against the upfront investment required, ensuring a smooth and scalable transformation.

8.8 Frontline Employee Involvement: A Key to Success

Finally, one of the most important aspects of a successful digital transformation is involving frontline employees early in the process. While the project team may design and implement the changes, it is the frontline workers who will use these systems daily. Their input can provide critical insights into potential pain points or areas for improvement that may not be immediately obvious to project managers or engineers.

By engaging frontline workers from the outset, the implementation can be more accurately tailored to not only meet corporate goals but also align with the practical, day-to-day needs of the workforce. This ensures that the new systems are user-friendly, efficient, and ultimately embraced by those who rely on them most.

8.8.1 The Importance of Frontline Feedback

For any digital transformation to truly succeed, it is essential that frontline employees provide candid feedback during the design phase. These employees, with their hands-on experience and intimate knowledge of day-to-day operations, offer invaluable insights that can significantly shape solutions to be both practical and user-friendly. Establishing feedback loops early in the process is crucial. Employees should be encouraged to share their thoughts on what works, what doesn't, and what could be improved. When their feedback is not only heard but acted upon, it increases the likelihood of developing a system that aligns with their needs, ultimately facilitating smoother adoption. This collaborative design approach ensures that the tools introduced are both beneficial and intuitive for those who use them most frequently.

8.8.2 Training for Successful Adoption

Training is a critical component of successful digital transformation. Comprehensive, hands-on training sessions are necessary to ensure that workers are comfortable using new tools and technologies. However, traditional one-time training sessions may fall short. Continuous learning and ongoing support are essential for reinforcing these skills and addressing issues as they arise.

Digital tools often lead to the creation of new job roles or responsibilities for frontline workers. For instance, the introduction of advanced data analytics platforms may necessitate that operators develop basic data interpretation skills. While this can be daunting, it also presents an opportunity for professional growth. Training programs should focus on both technical skills and soft skills—such as problem-solving and critical thinking—to help employees adapt to these new demands and boost their confidence in their roles.

8.8.3 Empowering Frontline Champions

It is crucial for frontline employees to become advocates for new technology. When employees support and champion the tools they use, it creates a positive ripple effect across the organization, facilitating easier and more effective adoption. Identifying and empowering "super users"—those particularly adept at using the new technologies—can be a key strategy. These individuals can serve as role models, leading by example and offering support to their peers during the adoption phase. By allowing super users to take on leadership roles, companies not only increase engagement but also foster a grassroots movement of support for the digital transformation.

8.8.4 Providing Ongoing Support

Frontline employees need ongoing support to ensure a smooth transition. In traditional manufacturing settings, workers typically know who to contact when something goes wrong. A similar level of targeted support must be available for digital applications. Instead of relying on generic IT help desks that may lack context, specialized support should be readily accessible. This ensures that technical issues are resolved quickly and effectively, reducing frustration and minimizing disruptions in operations. Additionally, empowering employees to make certain changes themselves, rather than waiting for the project team to intervene, fosters a sense of ownership and accountability.

8.8.5 Ensuring Security and Access Control

Security and access control are paramount in a digital transformation. It is essential to carefully manage who has access to what within digital systems. Ensuring that employees have the appropriate level of access—neither too much nor too little—protects both the technology and the workforce. Clear guidelines on access levels, coupled with training on security protocols, help safeguard sensitive information while allowing employees to perform their roles efficiently.

References

1. European Commission, Directorate-General for Research, Innovation, A. Renda, S. Schwaag Serger, D. Tataj, A. Morlet, D. Isaksson, F. Martins, M. Mir Roca, C. Hidalgo, A. Huang, S. Dixson-Declève, P.-A. Balland, F. Bria, C. Charveriat, K. Dunlop, and E. Giovannini. Industry 5.0, a transformative vision for Europe—Governing systemic transformations towards a sustainable industry. Publications Office of the European Union, 2021.
2. European Commission, Directorate-General for Research, and Innovation. ERA industrial technologies roadmap on human-centric research and innovation for the manufacturing sector. Publications Office of the European Union, 2024.
3. European Commission, Directorate-General for Research, Innovation, M. Breque, L. De Nul, and A. Petridis. Industry 5.0—Towards a sustainable, human-centric and resilient European industry. Publications Office of the European Union, 2021.
4. Saeid Nahavandi. Industry 5.0—a human-centric solution. *Sustainability*, 11(16), 2019.

Human Involvement and Roles 9

9.1 Automation and Augmentation for Frontline Employee

The digital transformation of manufacturing is reshaping the experience of frontline employees in profound ways. As automation and augmentation become integral components of modern manufacturing processes, their impact on employee roles, workflows, and decision-making cannot be overstated. Understanding the balance between these two concepts—automation and augmentation—is key to designing systems that empower frontline workers, drive operational efficiency, and enhance safety.

9.1.1 Automation: Improving Workflow Efficiency

Automation, in particular, is revolutionizing many of the repetitive tasks that have historically consumed significant amounts of time for frontline workers. According to the People + AI Research (PAIR) team at Google, automation can increase efficiency, improve safety, and enable new experiences [1]. For frontline employees who often deal with repetitive, manual tasks, the introduction of automation can bring about significant improvements in both productivity and work quality.

For instance, consider a typical manufacturing environment where material data must be logged and tracked at multiple stages of production. Traditionally, this would involve manual input, a process prone to human error and inefficiency. With automation, such tasks can be streamlined through the use of technologies like computer vision. A system equipped with computer vision can capture material information without requiring manual data entry, thus reducing errors and freeing up time for employees to focus on more complex or value-added tasks.

Additionally, automation plays a critical role in enhancing workplace safety. In environments where employees work in confined spaces or handle hazardous materials, it is often safer and more efficient to deploy sensors or robotics for monitoring rather than relying on human checks. For example, automation can detect overfilling in tanks and trigger automatic shut-off points, which helps prevent potentially catastrophic events. By providing real-time alerts and automated responses, digital systems can help employees make informed decisions and react more quickly to situations that could lead to accidents or production downtime.

One of the most profound benefits of automation for frontline employees is the centralization of data [1]. In traditional manufacturing setups, workers may need to collect data from various systems or log information in different locations, creating inefficiencies and potential for oversight. Automating these data-gathering processes and centralizing the information not only boosts efficiency but also empowers employees with a holistic view of the manufacturing process. By having access to comprehensive data, workers can gain insights into adjacent processes and develop a deeper understanding of how their role fits into the larger operation. This improved visibility not only enhances their expertise but also adds a sense of meaning and purpose to their work.

9.1.2 Augmentation: Empowering the Human Element in Manufacturing

While automation focuses on removing repetitive tasks from employees' workloads, augmentation takes a different approach. Rather than replacing human effort, augmentation provides tools and technologies that enhance workers' capabilities, enabling them to perform their jobs more effectively [1]. In a manufacturing context, augmentation is particularly valuable when it comes to problem-solving and drawing new insights from complex data streams.

Augmentation in manufacturing typically involves using digital tools and artificial intelligence (AI) to assist in decision-making processes. While AI systems are not yet advanced enough to solve all manufacturing challenges autonomously, they can provide critical information that allows employees to critically evaluate situations and determine the best course of action. For instance, real-time data streaming from multiple sources can be used to identify the root cause of a problem much more quickly than relying solely on human observation. The ability to collect data, receive feedback, and implement solutions in a shorter timeframe can significantly enhance overall productivity and reduce downtime in manufacturing operations.

One of the key advantages of augmentation is its ability to provide frontline employees with additional insights that they might not have been able to derive on their own. For example, in a scenario where a machine begins to show signs of malfunction, an augmented system could suggest possible causes and potential fixes based on real-time data and historical performance trends. This not only speeds up the troubleshooting process but also enables employees to make more informed decisions, reducing the likelihood of errors and improving overall operational efficiency.

For managers, augmentation tools can streamline the oversight of complex manufacturing processes. Managing a production line involves tracking numerous variables, and it is difficult for any individual to monitor everything in detail. However, with digital augmentation systems in place, managers can receive real-time data feeds that highlight key performance indicators, potential bottlenecks, or areas where intervention may be needed. These systems can also assist in scenario planning and modeling, enabling managers to predict the impact of certain decisions on the manufacturing process before they are implemented.

To fully realize the benefits of augmentation, it is essential to consider the design phase of digital transformation carefully. Designers must take into account how data will be used by both frontline employees and managers under various scenarios. This involves not only centralizing data collection but also ensuring that the information is easily accessible and can be exported for further analysis using familiar tools, such as Excel. While it may be tempting to build a comprehensive system that consolidates all functions into one platform, flexibility should be prioritized. Often, augmentation is most powerful when employees can quickly find the data they need and draw actionable insights from it.

9.1.3 Designing Systems that Balance Automation and Augmentation

Incorporating automation and augmentation into a manufacturing environment requires careful planning and consideration of the specific needs of the workforce. While automation can dramatically improve efficiency by handling repetitive tasks and enhancing safety, it is important to recognize that some processes still require human judgment and decision-making. Augmentation fills this gap by providing employees with the information and tools they need to make better decisions in real time.

During the design phase, it is critical to assess which tasks can be fully automated to achieve efficiency gains while leaving room for augmentation where human expertise is necessary. For example, while an automated system can monitor equipment health and flag potential issues, human employees still need to evaluate these alerts and determine the appropriate response. In this case, the role of augmentation is to ensure that employees have the right data at their fingertips and the insights needed to make the best possible decisions.

A thoughtful approach to balancing automation and augmentation can have a transformative effect on the experience of frontline employees. Automation can free them from mundane, repetitive tasks, while augmentation empowers them with the data and tools they need to perform their jobs more effectively. Together, these technologies enable employees to develop deeper expertise, make more informed decisions, and contribute to a safer, more efficient manufacturing environment.

9.1.4 The Human-Centric Future of Manufacturing

As manufacturing continues to evolve through digital transformation, the role of the frontline employee is shifting from one of manual labor to that of a data-informed decision maker. Automation and augmentation are at the heart of this shift, providing the tools and technologies needed to create a more human-centric manufacturing environment. By embracing these advancements, companies can not only improve their operational efficiency but also enhance the job satisfaction and overall experience of their employees.

Automation and augmentation are not mutually exclusive concepts; rather, they complement each other in driving the future of manufacturing. Automation is ideal for tasks that are repetitive, dangerous, or time-consuming, while augmentation is best suited for situations that require human insight and critical thinking. By thoughtfully integrating both into manufacturing processes, companies can unlock new levels of productivity, safety, and employee engagement.

9.2 Designing a System for Automation and Augmentation: A High-Level Guide

Creating an effective system that leverages both automation and augmentation in a manufacturing setting requires a thoughtful and structured approach. The balance between these two elements is key: while automation can streamline repetitive tasks, augmentation ensures that human expertise remains central to decision-making and problem-solving. Below is a high-level guide to designing such a system, ensuring that it maximizes operational efficiency while empowering frontline employees.

9.2.1 Define the Goals and Identify Use Cases

The first step in designing a system that incorporates automation and augmentation is to clearly define the goals you want to achieve. What specific problems are you trying to solve, and what outcomes are you aiming for? These goals should be centered around priorities such as efficiency, safety, and improving the employee experience.

Once the goals are defined, identify the key use cases where automation and augmentation can bring the most value. These could include repetitive tasks such as data entry or material handling for automation, and decision-making or problem-solving tasks for augmentation. For example, automating manual data logging through sensors and AI-powered vision systems can improve accuracy, while augmenting decision-making by integrating real-time data feeds for predictive maintenance.

9.2 Designing a System for Automation and Augmentation: A High-Level Guide

Key questions to consider:

- What are the pain points in the current process?
- Which tasks can be fully automated to improve efficiency and reduce human error?
- Where is human judgment critical, and how can it be enhanced with augmentation?

9.2.2 Map Current Workflows and Processes

Before implementing any system changes, it is essential to have a deep understanding of the existing workflows and processes. Create detailed process maps that outline the tasks performed by frontline employees, noting where data is collected, how decisions are made, and where bottlenecks or inefficiencies exist.

Mapping the current processes allows you to identify the specific areas where automation can eliminate unnecessary steps and where augmentation can provide better decision-making tools. These maps should also take into account cross-functional tasks where different teams or departments interact, ensuring that any changes are aligned across the entire operation.

Key considerations:

- Document how tasks are currently performed, including who performs them, how frequently, and with what tools.
- Identify any process overlaps or redundancies that could be streamlined through automation.
- Pinpoint decisions or complex tasks where human input is necessary and how augmentation could enhance that input.

9.2.3 Prioritize Data Centralization and Integration

One of the critical components of both automation and augmentation is data. Effective systems require centralized, integrated data streams that allow both automated processes and augmented decision-making to work seamlessly. Start by determining what data needs to be collected and how it will be used, ensuring that the system design includes the necessary sensors, monitoring tools, and real-time analytics.

Centralizing data in a single location provides a foundation for both automated responses and human decision-making. Automation thrives on clean, reliable data streams, while augmentation allows employees to draw insights from comprehensive datasets. This centralization enables smoother operations and better coordination between employees and machines.

Key actions:

- Implement tools that collect real-time data (e.g., sensors, IoT devices) and ensure they are compatible with existing systems.
- Use cloud-based or centralized databases to store and organize data, ensuring it is easily accessible for both automated processes and human analysis.
- Design dashboards or interfaces that present data in a user-friendly way for employees who will use augmented tools to make decisions.

9.2.4 Balance Automation and Human Judgment

Not all tasks can or should be automated. One of the most important principles in designing a system that combines automation and augmentation is to recognize where human judgment adds value. During the design phase, categorize tasks into those that can be automated without sacrificing quality and those where human input is essential.

For example, tasks such as monitoring equipment health can be largely automated with sensors, but decisions about how to respond to equipment failure should involve augmented tools that provide real-time data and suggestions, empowering employees to make the final call.

Best practices:

- Automate repetitive tasks where human error is likely, such as data collection or routine safety checks.
- Use augmentation tools to assist with tasks that require critical thinking, complex problem-solving, or decision-making.
- Ensure that the system provides humans with actionable data rather than overwhelming them with too much information.

9.2.5 Design User-Centered Interfaces and Tools

Both automation and augmentation systems should be designed with the end user in mind—frontline employees. Whether it's an automated system that requires minimal user input or an augmented system that offers data and insights, the interfaces and tools should be intuitive and user-friendly.

During the design phase, consider the existing skills of the workforce and how new tools can integrate into their daily workflows without causing disruption. Simplicity is key—employees should be able to interact with the system easily and feel empowered rather than overwhelmed. Training and onboarding should be provided to ensure they are comfortable with the new technology.

Design elements to focus on:

- User-friendly dashboards for augmented systems that clearly display data, alerts, and actionable insights.
- Interfaces that simplify automated tasks, minimizing the need for manual input while allowing for overrides or adjustments when necessary.
- Flexibility in tool design, allowing employees to export or analyze data using familiar systems (e.g., Excel) for greater control.

9.3 Designing Systems to Capture Continuous Feedback

In manufacturing facilities, especially those undergoing digital transformation, collecting feedback from frontline employees must be an ongoing, seamless process. Creating systems to capture this feedback is not just about deploying software but designing tools that integrate smoothly into employees' daily workflows. These systems need to be user-friendly and easily accessible to capture real-time, relevant data without adding extra burdens to operators, technicians, or managers.

9.3.1 Embedding Feedback in Digital Systems

For feedback to be truly effective, it needs to be embedded directly into the digital systems that frontline workers use. One practical example is to integrate comment and annotation sections within equipment monitoring dashboards. Suppose a worker notices a machine's sensor reporting abnormal data. In that case, they can immediately annotate the issue, providing their observations and concerns directly in the system. This can trigger an automated workflow for review by the maintenance team or engineers, expediting troubleshooting and resolutions.

Consider a manufacturing facility using an Industrial Internet of Things (IIoT) platform. Workers could leave real-time feedback, such as "Sensor A seems to be reporting unusually high temperatures compared to Sensor B," or "The predictive maintenance alert for this machine seems inaccurate." By embedding these features into daily-use systems, operators can leave feedback without interrupting their workflow. This also ensures that feedback is collected contextually, in real-time, when issues arise, which greatly enhances accuracy and actionability.

Such systems should be intuitive and require minimal additional effort to interact with. If an operator is on a production line, their primary focus is maintaining operational efficiency, not troubleshooting software. If providing feedback is difficult, workers are unlikely to engage. An example of an intuitive interface could be a tablet-based dashboard that allows quick annotation via touch input, similar to leaving a comment in a text box. This system could allow the worker to capture their observations directly where the issue occurs, in real-time, without navigating through multiple screens.

9.3.2 Evolving Systems Based on User Feedback

As Microsoft's guidelines for human-AI interaction highlight, systems should not be static; they must evolve based on user interactions and feedback [2]. Feedback mechanisms should continually improve to accommodate changes in the work environment or user needs. For example, if operators frequently suggest changes to the user interface because certain functions are difficult to use, the digital system should adapt and be updated based on this feedback.

A real-world example of this is the use of digital twins—virtual replicas of physical assets or processes in manufacturing. A digital twin can be continuously updated with real-time data from the physical system it mirrors, and operators can input feedback regarding discrepancies between the physical and digital environments. If a digital twin incorrectly models the behavior of a machine, operators can provide immediate feedback within the system, highlighting the inconsistency, which engineers can then address. This process creates a feedback loop where the digital twin evolves and becomes more accurate based on user input.

Another example can be found in predictive maintenance software. Feedback from operators can help fine-tune the predictive algorithms. For instance, if workers frequently report false alarms or missed alerts from the system, engineers can adjust the software's predictive model to better fit actual operating conditions. This adaptive approach ensures that the system becomes more accurate over time, improving its usefulness and relevance to the workers who rely on it.

9.3.3 Encouraging Continuous Engagement

To maintain continuous engagement from frontline employees, feedback mechanisms must not only be easy to use but also rewarding. Employees are more likely to participate if they see that their feedback is taken seriously and leads to tangible improvements. One way to encourage this is through transparency—showing how feedback has led to changes in the system. For example, implementing a feedback tracking system that allows employees to follow the progress of their suggestions could significantly boost participation. If a technician submits feedback about a faulty sensor, the system could notify them when their issue has been reviewed, resolved, and the sensor recalibrated.

Another strategy is to implement a form of recognition or incentive for useful feedback. This could be as simple as highlighting employees whose feedback led to important system improvements in a company newsletter or offering small rewards for those who actively participate in improving digital systems.

The feedback systems should also promote continuous dialogue between different teams—engineers, operators, and managers—allowing for a comprehensive understanding of the operational challenges and successes. For example, weekly or monthly review meetings that include feedback summaries could provide a structured environment for discussing key insights, prioritizing feedback, and making decisions on which suggestions to implement.

Incorporating these feedback mechanisms ensures not only that the technology evolves but also that the workplace culture becomes more collaborative This reflects industry best practices that encourage designing adaptive systems that evolve based on user needs and feedback, supporting continuous improvement and usability [2]. As the systems evolve to better fit the operational context, the workforce is more likely to engage deeply with the technology, driving both productivity and innovation.

9.4 Different Persona for Implementation

In a manufacturing setting, successful digital transformation hinges on involving the right people at the right stages of the process. Some organizations make the mistake of relying on the same experts throughout the entire project, but this approach can lead to gaps in execution and adoption. Instead, it's critical to identify different personas, individuals with unique traits and strengths, who are best suited for each phase of the transformation. At a high level, these personas (describe below) help guide the system's design, testing, and adoption, supporting a more user-friendly and widely accepted implementation..

9.4.1 Design Phase: Subject Matter Experts (SMEs)

In the design phase of digital transformation, SMEs are the backbone of the operation. These individuals are not only well-versed in the current processes but also have a deep understanding of the nuances of the production floor and broader operational environment. Their expertise helps to shape the system, ensuring it is functional and relevant to the everyday challenges of manufacturing.

SMEs are typically people who have spent years, if not decades, in the trenches, gaining a wealth of knowledge about the intricacies of how things run. Their experience allows them to identify potential roadblocks and offer solutions early in the design process. While technical teams might be experts in system design, the SMEs ensure that the technology actually supports and enhances operational efficiency.

Key Traits:

Deep Knowledge of Operations: SMEs are often the go-to individuals for insights into the manufacturing process. Their experience is invaluable in ensuring the system aligns with real-world applications.

Pragmatic Thinking: While they understand the theoretical possibilities of a digital system, SMEs remain grounded in practical concerns, balancing ideal solutions with realistic constraints.

Process-Driven Mindset: They think in terms of workflows and processes, ensuring that the system supports smooth and efficient operations from start to finish.

While it might be tempting to involve these experts in all phases of the project, their value is most pronounced during the design phase. Their primary focus is on crafting a system that solves the right problems, not necessarily on testing or promoting adoption later on.

9.4.2 Testing Phase: Detail-Oriented Evaluators

Once the design is in place, the next phase—testing—requires a very different set of skills. The testing phase, particularly User Acceptance Testing (UAT), needs individuals who thrive on diving deep into the details. These detail-oriented evaluators are essential for examining the system with a fine-tooth comb, ensuring that no issue, however small, goes unnoticed.

Digital transformation projects can sometimes overlook the importance of this persona, but selecting individuals who have both the patience and precision to test every aspect of the system is critical. These are the people who will spend hours navigating every corner of the software, replicating real-world scenarios to see if the system can handle them, and documenting bugs or inefficiencies.

Key Traits:

Attention to Detail: These individuals notice the small things, which can often lead to bigger issues down the line. They excel at finding gaps and errors that might be missed by others.

Patience for Repetitive Tasks: UAT can be tedious. Evaluators have to perform the same tasks repeatedly to test different scenarios, and they need the endurance to remain focused throughout.

Analytical Mindset: They can break down complex systems into manageable parts, testing each component rigorously and ensuring that it aligns with the overall system design.

By having these evaluators focus solely on testing, rather than splitting their attention across multiple phases, they ensure that the system is robust and ready for rollout.

9.4.3 Soft Launch: Early Adopters and Technology Enthusiasts

During the soft launch, the system transitions from testing to real-world use. At this stage, you need a different type of persona: the early adopters and technology enthusiasts. These individuals love exploring new technology, and they have an innate curiosity about how things work. Their enthusiasm is contagious, which makes them critical in both testing the system in real-world conditions and influencing their peers to embrace the new technology.

Early adopters are often the first to troubleshoot and provide feedback on how the system performs in actual working environments. They can be counted on to find creative ways to use the system, stretching its capabilities and identifying areas for improvement that may not have been obvious in the design or testing phases.

Their most valuable trait, however, is their ability to influence others. In any organization, there is often resistance to change. These technology enthusiasts serve as champions for the system, setting an example and helping their colleagues see the benefits of the new tools.

Key Traits:

Curiosity for New Technology: These individuals are eager to explore new tools and systems, often finding creative ways to enhance their workflow with the technology.
Resilience in Troubleshooting: They are quick to identify and work through any issues that arise, seeing problems as opportunities for improvement rather than setbacks.
Influence Among Peers: Early adopters have a strong presence within their teams. Their excitement and willingness to engage with the system can inspire others to follow suit, making them crucial in overcoming resistance to change.

By involving these individuals in the soft launch, you not only receive valuable feedback on system performance but also begin to build a culture of adoption and acceptance throughout the organization.

9.4.4 Leadership Champion: Advocates for Transformation

One of the most overlooked personas in digital transformation project is the leadership champion. This individual plays a pivotal role in ensuring the success of the initiative, not by directly engaging with the technology, but by driving communication and support across all levels of the organization.

The leadership champion is typically someone in a senior role who believes in the value of the transformation and is committed to its success. They act as a bridge between corporate leadership, strategy teams, and the frontline employees, ensuring that the transformation aligns with broader organizational goals while addressing the practical concerns of those on the ground.

The champion's role extends beyond mere advocacy. They must actively listen to feedback from the front lines and communicate that to decision-makers. They also need to address concerns and uncertainties among employees, providing reassurance that the digital transformation will benefit everyone involved. Their presence and confidence in the system foster trust, making it easier for employees to embrace the change.

Key Traits:

Strong Communication Skills: The leadership champion needs to be able to clearly articulate the goals and benefits of the digital transformation to employees at all levels.
Confidence and Influence: Employees are more likely to embrace the system if they see that leadership is fully invested and confident in its success.
Ability to Advocate and Lead: The champion not only advocates for the system internally but also ensures that feedback from employees is heard and acted upon by corporate leaders.

A leadership champion can make or break the digital transformation. Their involvement helps smooth the transition, address concerns, and maintain momentum throughout the process. They ensure that the system is seen as a positive change rather than a disruption.

In any digital transformation, success depends on engaging the right personas at different stages of the project. SMEs drive design, detail-oriented evaluators ensure thorough testing, early adopters help with the soft launch, and a leadership champion keeps everyone aligned and motivated. By selecting people based on their traits rather than their titles, organizations can optimize their digital transformation efforts and set themselves up for long-term success.

9.5 Conclusion

Successfully implementing digital solutions requires a deep understanding of the diverse human roles involved, with attention to the different personas that are needed to shape the system's design and adoption. Rather than treating the system as static, it's essential to recognize its dynamic relationship with users and continually center the human experience throughout the design process. By doing so, we ensure that technology not only meets functional needs but also empowers the people who interact with it, fostering a more sustainable and effective digital transformation.

References

1. https://pair.withgoogle.com/guidebook/. Google pair. people + ai guidebook, 2019.
2. Saleema Amershi, Dan Weld, Mihaela Vorvoreanu, Adam Fourney, Besmira Nushi, Penny Collisson, Jina Suh, Shamsi Iqbal, Paul Bennett, Kori Inkpen, Jaime Teevan, Ruth Kikin-Gil, and Eric Horvitz. Guidelines for human-ai interaction. In *CHI 2019*. ACM, May 2019. CHI 2019 Honorable Mention Award.

10 Digital Transformation and Cultural Transformation

10.1 The Human Factor in Digital Transformation-Thoughts

While technological advancements are undeniably crucial, the success of any digital initiative hinges on how it interacts with and influences the human element. These are some thoughts delving beyond the technological aspects, focusing on how culture, human behavior, and the interplay between corporate and plant-level realities shape the success or failure of digital transformations in this complex industry.

10.2 Defining Culture in a Manufacturing Context

Culture within a manufacturing setting, particularly in the complex domain of chemical processing, is a multifaceted phenomenon. It encompasses shared values, beliefs, assumptions, and behaviors that guide the actions and interactions of individuals within the organization. These cultural elements are deeply ingrained, often evolving over decades, and significantly influence how work is performed, decisions are made, and relationships are formed [1].

10.2.1 Divergence Between Corporate and Plant Levels

While corporate culture often reflects broader company values, such as innovation, sustainability, and customer focus, these values can often feel abstract and distant to those working on the front lines of production. At the corporate level, these values might be articulated in mission statements, communicated through executive speeches, and promoted through marketing campaigns. However, their direct impact on the daily work of a production operator in a chemical processing plant may not always be immediately apparent.

In contrast, plant-level culture is deeply rooted in the realities of daily operations. In chemical processing plants, with their significant capital investments and complex equipment, operational requirements and safety protocols often take precedence over broader corporate directives. This creates a unique subculture characterized by a paramount focus on safety, an unwavering emphasis on operational excellence, and a high value placed on practical skills and expertise.

Safety is not merely a value at manufacturing setting; it's a core tenet, permeating every aspect of plant operations. This translates into a relentless focus on adherence to procedures, meticulous risk assessments, and a pervasive culture of vigilance [2]. Many decisions and actions are evaluated through the lens of safety. This emphasis on safety extends beyond individual actions to encompass a collective responsibility for ensuring the well-being of all personnel. A strong safety culture is not merely about avoiding accidents; it's about creating a shared understanding of the potential hazards, fostering a sense of collective responsibility, and empowering employees to actively participate in identifying and mitigating risks. This can manifest in various ways, from regular safety audits and toolbox talks to peer-to-peer observations and near-miss reporting systems.

Furthermore, plant-level culture places a strong emphasis on quality and operational excellence. In a chemical manufacturing environment, where product quality and consistency are paramount, achieving and maintaining high-quality standards is critical for success. This translates into a relentless focus on process optimization, equipment maintenance, and continuous improvement. A culture of continuous improvement is fostered, where teams constantly strive to identify and eliminate bottlenecks, optimize resource utilization, and maximize production output while maintaining the highest quality standards. This may involve implementing lean manufacturing principles, conducting root cause analyses of production issues, and leveraging data analytics to identify areas for improvement.

Given the complexity of chemical processes, plant-level culture highly values practical skills, technical expertise, and the ability to troubleshoot and solve problems quickly. Experienced operators and technicians are highly respected for their knowledge and ability to navigate the intricacies of plant operations. This emphasis on practical expertise fosters a strong sense of collective knowledge and a culture of knowledge sharing among plant personnel. Senior operators often mentor junior colleagues, passing on their valuable experience and expertise through on-the-job training, informal discussions, and the sharing of best practices. This intergenerational transfer of knowledge ensures that critical skills and expertise are preserved within the plant, ensuring continued operational excellence and a skilled workforce for the future.

10.2.2 Corporate and Plant Artifacts

While corporate slogans, values, and mission statements may be displayed prominently throughout an organization, their interpretation and implementation can vary significantly

10.2 Defining Culture in a Manufacturing Context

between corporate headquarters and individual manufacturing sites [1]. At the corporate level, these statements often serve as broad, aspirational goals, aiming to establish a consistent brand image and define overarching organizational values. For example, a corporate value of "innovation" might be broadly defined as a commitment to continuous improvement and the development of new technologies.

However, at the plant level, these abstract concepts must be translated into concrete actions and behaviors that are relevant to the unique challenges and priorities of the specific manufacturing site [1]. A corporate emphasis on "innovation" might translate differently in a plant focused on high-volume production compared to a plant specializing in research and development. In the former, "innovation" might manifest as the implementation of lean manufacturing techniques to optimize existing processes, while in the latter, it might involve the development and testing of new product formulations or the adoption of cutting-edge automation technologies.

Furthermore, the artifacts of plant-level culture often reflect the unique challenges and priorities of the site more directly than corporate-level pronouncements. Safety slogans, prominently displayed and regularly reinforced, serve as constant reminders of the critical importance of safety in chemical manufacturing. These slogans, often concise and impactful, are not merely decorative elements; they are integral to the safety culture, serving as daily reminders of the shared commitment to safety. For example, slogans like "Safety First, Always" or "Safety is Everyone's Responsibility" can be seen prominently displayed throughout the plant, serving as constant reminders of the importance of safety in every aspect of plant operations. In contrast, while a corporate value statement might include a general commitment to employee safety, it may lack the specific and immediate impact of a plant-level safety slogan that is consistently reinforced through daily safety talks, toolbox meetings, and employee recognition programs.

Visual cues, such as diagrams, checklists, and safety signage, are ubiquitous in a safety-conscious manufacturing environment. These visual aids provide clear and concise instructions, ensuring that all personnel are aware of safety protocols and emergency procedures [2]. For instance, lock-out/tag-out procedures, critical for ensuring the safety of maintenance personnel, are often visually represented through step-by-step diagrams and checklists, ensuring that all personnel understand and adhere to these critical safety measures. These visual cues serve as constant reminders of safety protocols, minimizing the risk of human error and promoting a culture of proactive risk mitigation. In contrast, while corporate safety manuals may contain detailed procedures, their impact may be diminished if they are not effectively communicated and reinforced at the plant level through visual aids and on-the-job training.

10.3 Digital Transformation as a Culture Shaper

The culture at a manufacturing site will inevitably change during digital transformation because digital technologies force a rethinking of foundational workplace dynamics, from how work is executed to how authority is distributed and decisions are made. These changes are not superficial; they challenge the deeply rooted habits, values, and norms that have traditionally governed manufacturing environments.

At its core, digital transformation alters how work is performed by automating tasks and introducing tools like artificial intelligence, robotics, and data analytics. Historically, manufacturing cultures have been built around manual labor, hands-on problem-solving, and the accumulation of experiential knowledge over years of practice [3, 4]. With digital technologies taking over repetitive and routine tasks, the emphasis shifts from physical execution to intellectual oversight. Machine operators are no longer valued solely for their ability to perform tasks manually but for their capacity to monitor, troubleshoot, and optimize automated systems. This creates an unavoidable cultural shift: the skills and expertise that once defined excellence on the factory floor are being replaced or augmented by those rooted in data analysis, systems thinking, and predictive maintenance. Employees accustomed to traditional roles may feel their identity in the workplace is being eroded, leading to uncertainty, resistance, and a need for cultural adaptation to embrace this new model of work.

Digital transformation also forces a reevaluation of how information is accessed and used, which significantly impacts workplace culture. Manufacturing sites have long operated under hierarchies where information flowed through controlled channels, with decision-making often reliant on intuition and experience. The arrival of real-time data dashboards, sensor-driven insights, and advanced analytics democratizes information, making it readily available across all levels of the organization. While this empowers employees to make more informed, data-driven decisions, it also challenges traditional norms where authority and respect were tied to exclusive knowledge or tenure. Employees must now learn to trust and leverage data over instinct more, which can be a difficult adjustment, especially for those who have relied on years of hands-on expertise to guide their actions. The cultural change arises from a necessary shift in mindset: employees must move from being knowledge custodians to becoming active participants in a collaborative, data-enabled decision-making process. This shift often sparks tension as individuals reconcile the value of their experience with the growing dominance of technology-driven insights.

Moreover, digital technologies disrupt power structures and communication patterns, reshaping how authority is perceived and exercised. In traditional manufacturing settings, seniority and experience often dictated influence and decision-making power. With digital tools providing transparent access to information and enabling broader participation in decision-making, authority is redistributed. Employees at all levels can access real-time data and contribute insights, weakening the influence of individuals who once held power through information control. This redistribution of authority can provoke resistance, particularly among those who feel their leadership or expertise is being undermined. Additionally, the increased visibility and speed of communication enabled by digital tools challenge conventional methods of coordination, requiring employees to adopt new collaboration techniques and embrace more agile ways of working. This shift not only redefines how teams operate but also impacts how respect and credibility are earned within the workplace.

Finally, the emotional and psychological impact of these changes cannot be overstated. Digital transformation often introduces a sense of vulnerability and uncertainty as employees grapple with the fear of obsolescence and the pressure to acquire new skills [5]. This can lead to heightened anxiety and resistance, which in turn shapes the cultural response to transformation [5]. Successful adoption of digital technologies depends not only on providing the necessary tools and training but also on fostering a culture of openness, adaptability, and continuous learning. Employees need to feel supported in their transition, and the organization must actively work to align cultural values with the demands of the digital age.

Ultimately, the culture will change because digital transformation redefines what it means to be valuable, how decisions are made, and who holds power in the organization. These shifts touch every aspect of the workplace, from individual roles to collective behaviors, forcing a cultural evolution to align with a new digital reality. Without this cultural adaptation, even the most advanced technologies will struggle to achieve their full potential in the manufacturing environment.

10.4 Building Norms Rather than Isolated Cultural Elements

True cultural transformation in manufacturing demands a grassroots approach, not top-down decrees from corporate HQ. Successful digital transformation hinges on the active participation and ownership of plant personnel. Plant-level teams possess the unique insights and understanding to craft and implement norms that perfectly align with their specific operational realities and challenges.

This necessitates fostering a sense of ownership and empowerment among plant personnel, encouraging their active involvement in the design and implementation of digital solutions.

10.4.1 The Peril of Isolated Cultural Events

A common pitfall in some digital transformation initiatives is the reliance on isolated cultural events as the primary driver of change. These events, such as workshops, training sessions, or even company-wide retreats, often lack the sustained impact necessary for true cultural transformation. While these events can be valuable for raising awareness, sharing best practices, and fostering initial enthusiasm, they rarely translate into lasting changes in everyday work habits and behaviors.

The problem lies in their inherent isolation. These events are often treated as one-off occurrences, disconnected from the day-to-day realities of the manufacturing floor. They fail to address the root causes of existing cultural challenges and provide insufficient support for employees to translate newly acquired knowledge and skills into tangible, ongoing improvements. As a result, the impact of these events tends to be fleeting, with employees quickly reverting to their old ways of working once the excitement of the event has subsided.

Furthermore, a sole focus on isolated cultural events can create a false sense of progress. By simply checking off a list of events, organizations may mistakenly believe they are successfully driving cultural change. This can lead to complacency and a neglect of more fundamental and sustainable approaches to transformation.

10.4.2 Building New Norms Through Digital Transformation

Instead of relying solely on isolated cultural events, organizations should leverage digital tools to foster the emergence of new, enduring norms that permeate the entire organization. Digital technologies can serve as powerful catalysts for cultural change by:

Revolutionizing Communication: Digital platforms can break down traditional communication silos and facilitate seamless information flow across departments and teams. Instant messaging platforms, project management tools, and collaborative workspaces can enable real-time communication, foster rapid response times, and promote a culture of transparency and open dialogue.

> *For example, a manufacturing plant could implement a company-wide instant messaging platform to facilitate quick and efficient communication between production teams, maintenance crews, and quality control departments. This can significantly reduce response times to urgent issues, improve coordination between teams, and minimize costly production delays.*

Driving Data-Driven Decision Making: Digital tools can empower plant personnel to make more informed and data-driven decisions. By leveraging data analytics and visualization tools, organizations can gather and analyze real-time data on production performance, equipment utilization, and quality metrics. This data can be used to identify areas for improvement, optimize production processes, and proactively address potential issues.

> *For instance, a plant could implement a sensor-based system to monitor equipment performance in real-time. This data can be analyzed to predict potential equipment failures, schedule preventive maintenance, and minimize costly downtime. This shift towards data-driven decision making can foster a culture of continuous improvement and empower plant personnel to take ownership of their work and drive positive change.*

Fostering Collaboration and Knowledge Sharing: Digital tools can facilitate seamless collaboration across departments and teams, breaking down traditional barriers and fostering a more integrated approach to problem-solving. Cloud-based platforms can enable remote access to data, documents, and project files, allowing teams to collaborate effectively regardless of their location.

> *For example, a plant could utilize a cloud-based platform to share best practices, troubleshoot equipment issues, and collaborate on process improvement initiatives. This can foster a sense of shared ownership and responsibility, encourage knowledge sharing across the organization, and accelerate the pace of innovation.*

Empowering Plant Personnel: By providing plant personnel with access to the right digital tools and training, organizations can empower them to take ownership of their work, drive innovation, and contribute to the overall success of the organization. This can foster a sense of pride and ownership among the workforce, leading to increased engagement, improved productivity, and greater job satisfaction.

10.5 Recognizing and Addressing Cultural Resistance

Digital transformation within manufacturing presents significant challenges, with cultural resistance often being a primary obstacle. One major hurdle arises from the inherent disconnect between corporate directives and the unique realities faced by individual manufacturing plants. Top-down mandates, often developed without adequate consideration for plant-specific needs and constraints, can be perceived as unrealistic or even detrimental to operations. This misalignment fosters a sense of disconnect and resentment among plant personnel, leading to passive resistance, active sabotage, or simply a lack of engagement with the initiative [6].

Furthermore, the introduction of automation and AI technologies, such as robotics and machine learning, inevitably raises concerns about job security and potential displacement. These anxieties, compounded by a lack of clear communication and a dearth of focus on employee training and development, can create a climate of fear and uncertainty within the workforce. Employees may fear that their existing skills are becoming obsolete, leading to a decline in morale, increased absenteeism, and even voluntary turnover. This resistance can manifest in various ways, from outright rejection of new technologies to subtle forms of sabotage, such as intentional data entry errors or a reluctance to fully utilize new systems.

The disruption of established routines and work habits also presents a significant challenge. Even positive changes can be unsettling, as the introduction of new technologies and processes inevitably disrupts existing workflows, communication patterns, and social dynamics within the plant. This can create anxiety and resistance among employees who may feel overwhelmed by the need to adapt to new ways of working. This resistance can manifest in various ways, such as procrastination, avoidance, and a general reluctance to embrace new technologies. Employees may resist training, sabotage new equipment, or simply continue to rely on old, inefficient methods.

Finally, a lack of employee involvement and ownership can significantly hinder the success of any digital transformation initiative. When employees feel excluded from the decision-making process and lack a sense of ownership over the initiative, resistance is more likely to arise. Top-down implementation, where decisions are made without input from plant personnel, can breed resentment and undermine trust. This lack of involvement can lead to a sense of disenfranchisement and a feeling that their concerns and perspectives are being ignored. This can result in passive resistance, lack of engagement, and ultimately, the failure of the initiative.

Addressing these challenges requires a multifaceted approach. Open and transparent communication is crucial, ensuring that employees are kept informed about the goals, benefits, and potential challenges of the digital transformation initiative. Addressing employee concerns proactively and transparently can help to mitigate resistance and build trust. Investing in comprehensive training and development programs is essential to equip employees with the necessary skills to thrive in the digital age. Providing opportunities for continuous learning and skill development can help to alleviate fears of job displacement and empower employees to embrace new technologies.

Furthermore, actively involving plant personnel in the decision-making process is critical. By creating opportunities for employees to provide feedback and input on the initiative, organizations can ensure that their concerns and perspectives are taken into account. This can help to build a sense of ownership and shared responsibility for the success of the initiative. Finally, implementing digital transformation initiatives in a phased manner, starting with pilot programs, can help to minimize disruption and allow for gradual adaptation. This approach allows for continuous feedback and refinement of the implementation process, addressing any emerging challenges or concerns promptly.

By addressing these challenges proactively and fostering a culture of open communication, collaboration, and continuous improvement, organizations can overcome cultural resistance and ensure the successful implementation of digital transformation initiatives within their manufacturing settings.

10.6 Fostering Cultural Shifts Through Digitalization

10.6.1 Increased Observability and Decreased Plausible Deniability

The introduction of digital technologies in manufacturing significantly increases observability and reduces plausible deniability, as noted by Andrew McAfee [7]. In a plant environment, where outcomes and processes are increasingly measurable in real-time through sensors, data logs, and other digital tools, employees become more accountable for their actions and decisions.

Real-time data provides unprecedented insights into individual and team performance, enabling more accurate and objective performance evaluations. For instance, sensors can track machine utilization, production output, and energy consumption, while data logs can record every step of a manufacturing process. This level of granularity allows managers to identify inefficiencies, pinpoint bottlenecks, and accurately assess individual contributions. With increased observability, it becomes more difficult for individuals or teams to avoid responsibility for their actions or to make excuses for poor performance. For example, if a machine consistently underperforms, it becomes harder to blame unforeseen circumstances when real-time data reveals patterns of neglect or improper maintenance. This increased transparency can lead to increased accountability and a greater sense of responsibility among employees, as they become more aware of the impact of their actions on overall plant performance.

This shift in accountability can create both opportunities and challenges. While it can incentivize improved performance and encourage a greater sense of responsibility, it can also create increased pressure and anxiety among employees. The increased observability brought about by digital technologies in manufacturing, while offering numerous benefits, also presents significant challenges. One major concern is the potential for increased pressure and anxiety among employees. The constant scrutiny of their work, even if intended to be constructive, can create a climate of fear and a sense of being constantly monitored. This constant pressure can stifle creativity and innovation, as employees may prioritize avoiding errors over taking risks and exploring new approaches.

Moreover, the sheer volume of data generated by modern manufacturing systems can be overwhelming. Analyzing and interpreting this data requires sophisticated analytical tools and skilled personnel. A lack of proper data management and analysis capabilities can lead to data overload, hindering decision-making and potentially even creating more confusion than clarity.

Finally, the rapid pace of technological change can create a constant sense of uncertainty and anxiety among employees. As new technologies are introduced, employees may feel overwhelmed by the need to constantly learn and adapt. This can lead to a feeling of inadequacy and a fear of being left behind, further exacerbating resistance to change.

The constant scrutiny of their work, even if it is intended to be constructive, can lead to feelings of stress and a fear of failure. Employees may feel constantly monitored, which can stifle creativity and innovation. Furthermore, the potential for misuse of data, such as unfair or biased performance evaluations, raises concerns about employee privacy and well-being. It is crucial to address these concerns proactively and ensure that data is used ethically and responsibly. Transparent communication about how data is collected, used, and shared is essential to build trust and mitigate concerns about privacy and surveillance. Clear guidelines and policies regarding data usage and employee privacy must be established and communicated to all employees. Moreover, it is important to emphasize that the primary goal of increased observability is not to create a culture of fear and blame, but rather to provide valuable insights for continuous improvement and employee development.

By focusing on the constructive use of data to identify areas for improvement, provide targeted training and support, and foster a culture of open communication and collaboration, organizations can leverage the benefits of increased observability while mitigating the potential downsides. This requires a shift in mindset from simply tracking performance to using data to empower employees, identify areas for growth, and create a more productive and fulfilling work environment.

10.6.2 Using Data to Challenge Bias and Drive Decision-Making

In manufacturing plant environments, long-standing biases, such as "we've always done it this way," can hinder innovation and limit progress. By leveraging data and analytics, plant personnel can challenge entrenched biases and make more informed and objective decisions. Data can help to identify inefficiencies and bottlenecks in production processes that may have been overlooked due to ingrained assumptions or biases. By continuously analyzing data and identifying areas for improvement, plants can break away from traditional, less efficient methods and embrace more innovative and data-driven approaches. Data can serve as a common ground for discussion and debate, providing an objective basis for evaluating different options and making informed decisions.

However, it's crucial to remember that data and analytics are not a replacement for human expertise and experience. While digital tools can provide valuable insights and identify potential areas for improvement, they should be used in conjunction with the deep knowledge and experience of plant personnel. Experienced operators, engineers, and chemists possess a wealth of knowledge about the intricacies of the manufacturing process, including nuances that may not be fully captured by data. Their insights are essential for interpreting data, identifying potential limitations, and developing effective solutions.

It is important to acknowledge that sometimes these "long-standing biases" may actually represent valuable knowledge and proven best practices developed over years of experience. Dismissing these experiences solely based on data-driven analysis can lead to unintended consequences and undermine the value of the collective expertise within the organization. The key is to find the right balance between challenging outdated assumptions and respecting the valuable knowledge and experience of plant personnel. Digital transformation should be a collaborative process that leverages both data-driven insights and the collective wisdom of the workforce.

10.6.3 Encouraging a Culture of Scientific Debate and Psychological Safety

Establishing a culture where data-backed arguments are encouraged and rewarded is paramount for driving continuous improvement within a manufacturing facility especially in the digital transformation [7]. This requires a fundamental shift in mindset, moving away from a culture of unquestioned authority and towards one of open inquiry and collaborative decision-making especially when data is plentiful and democratized. Employees should feel comfortable presenting conflicting viewpoints, knowing that their opinions will be not only considered but actively sought after. Creating this type of environment necessitates a psychologically safe space where individuals feel comfortable expressing their ideas and challenging assumptions without fear of reprisal [8]. This requires fostering a culture of open communication, active listening, and mutual respect. Managers and supervisors play a crucial role in cultivating this environment by actively encouraging open dialogue, valuing diverse perspectives, and creating opportunities for employees to share their insights and concerns.

Recognizing and rewarding contributions to constructive debates can further incentivize open dialogue and encourage a culture of continuous learning and improvement. This could include acknowledging individuals who effectively present data-driven arguments, provide valuable insights, or challenge existing assumptions in a constructive manner. By publicly recognizing these contributions, organizations can demonstrate the value they place on open communication and critical thinking. Over time, these ongoing debates create a powerful dynamic, establishing norms of openness, curiosity, and continuous improvement. This leads to a culture where learning and innovation are prioritized over rigid adherence to established practices. By encouraging open and honest discussions, organizations can foster a culture of continuous learning and adaptation, enabling them to respond effectively to the ever-changing demands of the market, technological advancements, and increasing competition.

Addressing the fear of failure is crucial. By creating a culture where mistakes are viewed as learning opportunities rather than personal failures, employees will be more willing to take risks and challenge the status quo [8]. Building trust between employees and management is essential for fostering open communication. This requires consistent and transparent communication, fair and equitable treatment of all employees, and a commitment to addressing concerns and resolving conflicts constructively. Digital tools can play a significant role in facilitating open communication and knowledge sharing. Platforms for collaborative work, online forums, and data visualization tools can empower employees to share information, exchange ideas, and engage in productive discussions. Investing in the professional development of employees is crucial for cultivating a data-driven culture. This includes providing training on data analysis, critical thinking, and effective communication skills. By cultivating a culture that values open dialogue, embraces data-driven decision-making, and encourages continuous learning, manufacturing organizations can unlock their full potential, enhance their competitiveness, and thrive in an increasingly dynamic and challenging environment.

10.6.4 Shifting Norms Through Data-Driven Discussions

Engaging in data-driven debates not only improves decision-making but also gradually shifts the underlying culture of the plant. By engaging in data-driven discussions, plant personnel can gradually shift away from relying solely on personal experience and intuition and towards a more data-driven approach to decision-making. As data-driven decision-making becomes more commonplace, new norms can emerge, emphasizing the importance of evidence-based decision-making, critical thinking, and a willingness to challenge assumptions. Over time, these shifts in decision-making practices can lead to a gradual transformation of the plant culture, fostering a more data-driven, analytical, and innovative environment.

10.6.5 Promoting Collaboration Through Data Sharing

Siloed operations in chemical plants significantly impede collaboration and hinder overall efficiency. When departments operate in isolation, critical information remains confined within their own boundaries, leading to a fragmented understanding of plant operations. This lack of information sharing can result in duplicated efforts, misaligned priorities, and a suboptimal allocation of resources. For example, the maintenance team might be unaware of upcoming production schedules, leading to unplanned equipment downtime and costly production delays. Conversely, the production team might not be fully informed about planned maintenance activities, potentially disrupting production schedules and impacting overall plant output.

Digital transformation offers a powerful solution to overcome these limitations. By implementing interconnected systems and platforms, such as Manufacturing Execution Systems (MES), Enterprise Resource Planning (ERP) systems, and cloud-based data warehouses, chemical plants can facilitate seamless data sharing across departments. This increased data visibility allows teams to gain a holistic understanding of plant operations and identify areas for improvement.

Furthermore, digital technologies enable real-time data sharing and collaboration, fostering a more responsive and agile operating environment. This facilitates faster decision-making and improved coordination across departments. For instance, real-time data from sensors and control systems can be analyzed to identify emerging issues, such as equipment malfunctions or deviations from expected production parameters. This information can be immediately shared with relevant teams, enabling them to take corrective action quickly and minimize the impact on overall plant operations. By leveraging digital technologies to break down data silos and promote data sharing, chemical plants can create a more interconnected and collaborative operating environment. This not only enhances operational efficiency and reduces downtime but also fosters a culture of continuous improvement, where teams can work together to identify and address challenges, optimize processes, and drive innovation.

10.6.6 Using Digital Tools to Train and Build Capabilities

In the competitive landscape of modern manufacturing, high employee turnover presents a significant challenge to organizational success. The loss of experienced personnel inevitably leads to a critical loss of institutional knowledge, severely impacting operational efficiency, productivity, and overall plant performance. When seasoned employees depart, they take with them a wealth of invaluable information: intricate process knowledge gleaned from years of hands-on experience, troubleshooting techniques honed through countless encounters with unexpected challenges, and best practices refined through continuous experimentation and adaptation. This loss of tacit knowledge, the deeply ingrained and often unspoken expertise that resides within individuals, can have profound consequences.

For instance, the departure of a skilled maintenance technician can leave a void in critical knowledge areas, such as the nuances of specific equipment, the history of past maintenance issues, and the most effective troubleshooting methods for recurring problems. This knowledge gap can significantly impact the ability of remaining staff to effectively maintain equipment, leading to increased downtime, reduced equipment lifespan, and potentially even safety hazards. Similarly, the loss of a seasoned production supervisor can disrupt production flow, as their in-depth understanding of production bottlenecks, process optimization techniques, and the subtle cues that indicate potential problems are irreplaceable. This can lead to decreased productivity, increased waste, and a decline in product quality.

To mitigate this critical knowledge drain, manufacturing organizations must embrace digital solutions to capture, preserve, and disseminate institutional knowledge. Online training platforms offer a robust foundation for addressing this challenge. These platforms can provide access to a comprehensive library of training materials, including high-quality videos demonstrating complex procedures, interactive simulations that replicate real-world scenarios, and engaging exercises that reinforce key concepts. For example, virtual reality (VR) simulations can replicate the operation of complex machinery, allowing trainees to practice tasks such as equipment startup, shutdown, and troubleshooting in a safe and controlled environment before attempting them on the actual production line. This immersive experience provides valuable hands-on experience and enhances knowledge retention significantly compared to traditional classroom-based training. Furthermore, these platforms can incorporate gamification elements, such as leaderboards and rewards, to increase engagement and motivation among trainees.

Beyond formal training programs, digital platforms can be leveraged to create and maintain a central repository of knowledge, serving as a digital encyclopedia of best practices, troubleshooting guides, and critical operational information. This centralized knowledge base, readily accessible to all employees, ensures that critical information remains readily available and easily searchable. For example, a plant-wide knowledge management system can store and index a vast array of documents, including standard operating procedures (SOPs), maintenance logs, technical manuals, and even expert interviews with retiring employees. This digitized knowledge base can be easily searched and filtered, allowing employees to quickly find the information they need, when they need it.

Furthermore, digital tools can facilitate seamless knowledge sharing and collaboration among employees. Online forums, discussion groups, and knowledge-sharing platforms can encourage the exchange of ideas, the sharing of best practices, and the resolution of complex challenges through collective expertise. Experienced technicians can share their troubleshooting tips and best practices with newer employees, while younger employees can bring fresh perspectives and innovative ideas to the table. This collaborative environment fosters a culture of continuous learning and knowledge sharing, ensuring that the collective knowledge of the workforce is effectively utilized and preserved. For instance, a plant-wide online platform could be established where employees can document and share successful problem-solving strategies, record lessons learned from past incidents, and discuss best practices for improving specific processes. This platform would serve as a living repository of collective knowledge, constantly evolving and growing as employees contribute their insights and experiences.

By effectively leveraging digital technologies for training, knowledge management, and collaboration, manufacturing organizations can mitigate the significant challenges associated with high employee turnover. By creating a robust digital infrastructure for knowledge capture, dissemination, and sharing, organizations can ensure that critical knowledge is preserved, employee skills are continuously developed, and the operational memory of the plant is maintained. This not only improves operational efficiency, productivity, and safety but also enhances the organization's ability to adapt to changing market demands, embrace new technologies, and remain competitive in the increasingly dynamic and challenging global marketplace.

10.6.7 Enhancing Corporate Visibility with Digital Data

Digital tools provide regional and corporate offices with more granular insights into plant operations. Instead of relying on aggregated metrics and KPIs, corporate leadership can gain a clearer understanding of on-the-ground realities. This helps align corporate strategies with plant-level operations, ensuring that decisions are based on real-time, actionable data. For example, real-time data on production output, equipment performance, and safety incidents can provide valuable insights into plant-level operations, allowing corporate leadership to make more informed decisions about resource allocation, investment priorities, and overall business strategy.

Digital systems facilitate communication between plant and corporate levels, promoting a better understanding of each other's challenges and capabilities, and fostering a more cohesive culture across the organization. Video conferencing, instant messaging, and collaborative workspaces can enable more frequent and effective communication between plant personnel and corporate teams. This can help to break down communication silos and improve coordination and collaboration across the organization.

10.7 Conclusion

In chemical processing industries, culture is not static; it evolves with the introduction of new technologies, the changing workforce, and the challenges of modern manufacturing. Digital transformation plays a critical role in this evolution by reshaping how culture is experienced, enhancing decision-making, promoting collaboration, and addressing resistance.

Digital transformation plays a critical role in this evolution by reshaping how culture is experienced, enhancing decision-making, promoting collaboration, and addressing resistance. By intentionally managing the intersection of culture and digital tools, chemical plants can create a more agile, data-driven, and collaborative environment. This involves fostering a culture of continuous learning and improvement, encouraging open dialogue and debate, and addressing employee concerns proactively.

References

1. E.H. Schein and P.A. Schein. *Organizational Culture and Leadership*. The Jossey-Bass Business & Management Series. Wiley, 2016.
2. Hervé Laroche, Corinne Bier, Claude Gilbert, and Benoît Cham Journé, editors. *Safety Cultures, Safety Models: Taking Stock and Moving Forward*. Springer International Publishing, 2018.
3. Emma Griffin. *Liberty's Dawn: A People's History of the Industrial Revolution*. Yale University Press, 2014.
4. Vaclav Smil. *Creating the Twentieth Century: Technical Innovations of 1867-1914 and Their Lasting Impact*. Oxford University Press, 2005.
5. D. Plekhanov, H. Franke, and T. H. Netland. Digital transformation: A review and research agenda. *European Management Journal*, 41(6):821–844, 2023.
6. Richard Dunford, Ian Palmer, and David Buchanan. *Managing Organizational Change: A Multiple Perspectives Approach*. McGraw-Hill Education, 2016.
7. Andrew McAfee and Erik Brynjolfsson. *A Race Against the Machine: How the Digital Revolution is Accelerating Innovation, Driving Productivity, and Irreversibly Transforming Employment and the Economy*. Digital Frontier Press, Lexington, MA, 2011.
8. Amy C Edmondson. *Right kind of wrong: The science of failing well*. Simon and Schuster, 2023.

Epilogue–Outlook   11

11.1 Fundamental Strategic Goals

The adoption of digital transformation in process industry operations is expected to grow significantly over the next decade. Digitalization enables industries to make complex decisions in a transparent and novel manner, revolutionizing supply chains, production, and business models.

Many existing plants were not designed with digitalization, autonomous optimization, or remote control in mind. To successfully implement digital transformation, the following strategic goals should be considered.

11.1.1 Experimentation and Evaluation

Evaluating the impact of digital technologies on plant operations is crucial. Establishing proper metrics and conducting case studies can help determine their benefits. Experiments should ideally be performed on relevant plants to avoid inefficiencies from adopting solutions that do not fit the process. Blind replication of digital implementations from one plant to another can lead to unsatisfactory outcomes.

11.1.2 Building an Ecosystem

Assessing an organization's digital maturity helps define feasible digitalization strategies. Organizations must determine which technological capabilities can be built internally and which require external expertise.

Digital transformation should align with the overall business automation framework. Understanding networking strategies, resource planning, information flow, organizational structure, and security management ensures a seamless integration of plant digitalization into the business informatics framework.

11.1.3 Scaling at the Edges

A phased approach, beginning with smaller-scale implementations, allows organizations to test and refine strategies with minimal risk. Chemical industries can initiate digitalization in auxiliary processes such as water management, utilities, and inventory systems before scaling up to core production processes. Full-scale digitalization is a significant undertaking and must be approached carefully to avoid disruptions.

11.1.4 Proving with Strategic Implementations

Prioritizing areas with high potential value fosters sustainable growth. Digitalization provides varied benefits across processes; hence, assessing implementations through diverse performance metrics is essential.

Success metrics must integrate perspectives from different stakeholders. Digitalization should enhance decision-making, predictive capabilities, and operational efficiency, impacting energy savings, uptime, maintenance, and risk mitigation.

11.1.5 Iterative Improvement

With rapidly evolving digital technologies, continuous refinement of digitalization efforts is necessary. Brownfield plants offer opportunities to test and optimize digital solutions, creating valuable operational databases.

These efforts contribute to smarter plant designs by leveraging insights on seasonal variations, raw material conditions, and operational efficiency. Digitalization not only optimizes existing plants but also lays the foundation for future smart plants through iterative innovation.

11.2 Common Misconceptions and Hyper-Expectations

Digitalization in the industrial sector is often met with misconceptions, apprehensions, and unrealistic expectations that may hinder its effective implementation. While automation, data-driven decision-making, and advanced simulations have significantly improved efficiency, understanding their limitations is crucial.

Digitalization Eliminates Process Uncertainty: A common assumption is that digital models provide absolute certainty about process behavior. However, industrial processes are influenced by stochastic factors such as feed composition variations, environmental conditions, and equipment aging. While digital twins and simulations enhance predictability, they do not eliminate inherent uncertainties.

AI can Replace Fundamental Process Knowledge: The increasing use of AI in industrial applications has led to the belief that machine learning models can substitute for engineering principles. In reality, data-driven models require robust domain knowledge for validation, calibration, and interpretation. AI complements traditional engineering by identifying patterns, but process understanding remains indispensable.

Digitalization Eliminates the Need for Human Operators: Another misconception is that automation can entirely replace human decision-making. While control systems optimize operations, industrial environments require human expertise for troubleshooting, strategic planning, and adapting to unexpected conditions. Digitalization enhances decision support rather than replacing operators.

Addressing these misconceptions requires a multidisciplinary approach. Digitalization is a powerful tool for industrial advancement, but its limitations must be understood. Recognizing the role of uncertainty, the necessity of process knowledge, and the importance of human operators ensures a balanced and effective approach to digital transformation. Training programs, cross-functional teams, and collaboration between various organizational denizens with diverse backgrounds are essential for a successful implementation and adoption of digital transformation.

11.3 Shaping the Future

The digital revolution in chemical processing industries is not just about replacing technology, it is about fundamentally transforming the way we manufacture. When done right, digitalization has the power to embed sustainability into every aspect of operations, drive higher efficiency and resilience, and, most importantly, create a more human-centric approach that benefits both employees and customers. This is not simply an upgrade; it is a reimagining of how businesses operate, making factories smarter, more agile, and more adaptable to future challenges. More than just improving productivity, digitalization enables a more responsive and innovative manufacturing ecosystem, ensuring enterprises stay competitive while making a positive impact on the world.

Yet, for all its potential, digitalization has often been narrowly focused on upgrading machinery, processes, and plants, overlooking the critical role of people. The future of manufacturing will not be defined by technology alone, but by how well humans and machines work together. As factories become more data-driven, the role of employees will evolve from routine monitoring to strategic problem-solving, leveraging digital tools to predict, prevent, and optimize operations. A successful transformation ensures that digitalization enhances—not replaces—human expertise, creating workplaces that are more engaging, intellectually stimulating, and aligned with a shared vision of progress. Companies that recognize this shift and invest in both technology and people will lead the way in shaping a future that is not only efficient and resilient but also fundamentally human at its core.

The manufacturer's authorised representative in the EU is Springer Nature Customer Service Centre GmbH, Europaplatz 3, 69115 Heidelberg, Germany. If you have any concerns regarding our products, please contact ProductSafety@springernature.com

Printed and bound by CPI Group (UK) Ltd, Croydon, CR0 4YY
26/03/2026
02078991-0004